School Bus and Truck Collision at Intersection
Near Chesterfield, New Jersey
February 16, 2012

I0482680

Accident Report

NTSB/HAR-13/01
PB2013-106638

National
Transportation
Safety Board

NTSB/HAR-13/01
PB2013-106638
Notation 8402A
Adopted July 23, 2013

Highway Accident Report

School Bus and Truck Collision at Intersection
Near Chesterfield, New Jersey
February 16, 2012

**National
Transportation
Safety Board**

490 L'Enfant Plaza, SW
Washington, DC 20594

National Transportation Safety Board. 2013. *School Bus and Truck Collision at Intersection Near Chesterfield, New Jersey, February 16, 2012.* Highway Accident Report NTSB/HAR-13/01. Washington, DC.

Abstract: On February 16, 2012, about 8:15 a.m., near Chesterfield, New Jersey, a Garden State Transport Corporation 2012 IC Bus, LLC, school bus was transporting 25 students to Chesterfield Elementary School. The bus was traveling north on Burlington County Road (BCR) 660, while a Herman's Trucking Inc. 2004 Mack roll-off truck with a fully loaded dump container was traveling east on BCR 528, approaching the intersection. The bus driver had stopped at the flashing red traffic beacon and STOP sign. As the bus pulled away from the white stop line and entered the intersection, it failed to yield to the truck and was struck behind the left rear axle. The bus rotated nearly 180 degrees and subsequently struck a traffic beacon support pole. One bus passenger was killed. Five bus passengers sustained serious injuries, 10 passengers and the bus driver received minor injuries, and nine passengers and the truck driver were uninjured. Major safety issues identified in this investigation were school bus driver fatigue, sedating prescription medications, medical conditions, and commercial driver's license medical examinations; truck driver speed, oversight of overweight commercial vehicles, brake maintenance, and final stage manufacturing air brake system installation; connected vehicle technology; and school bus occupant injuries and school bus crashworthiness. The National Transportation Safety Board makes recommendations to the Federal Motor Carrier Safety Administration; National Highway Traffic Safety Administration; states of California, Florida, Louisiana, New Jersey, New York, and Texas; National Truck Equipment Association; National Association of State Directors of Pupil Transportation Services; National Association for Pupil Transportation; National School Transportation Association; School Bus Manufacturers Technical Council; National Safety Council, School Transportation Section; and Herman's Trucking Inc.

The National Transportation Safety Board (NTSB) is an independent federal agency dedicated to promoting aviation, railroad, highway, marine, and pipeline safety. Established in 1967, the agency is mandated by Congress through the Independent Safety Board Act of 1974 to investigate transportation accidents, determine the probable causes of the accidents, issue safety recommendations, study transportation safety issues, and evaluate the safety effectiveness of government agencies involved in transportation. The NTSB makes public its actions and decisions through accident reports, safety studies, special investigation reports, safety recommendations, and statistical reviews.

Recent publications are available in their entirety at www.ntsb.gov. Other information may be obtained from the website or by contacting:

National Transportation Safety Board
Records Management Division, CIO-40
490 L'Enfant Plaza, SW
Washington, DC 20594
(800) 877-6799 or (202) 314-6551

Copies of NTSB publications may be purchased from the National Technical Information Service. To purchase this publication, order report number PB2013-106638 from:

National Technical Information Service
5301 Shawnee Road
Alexandria, VA 22312
(800) 553-6847 or (703) 605-6000

The Independent Safety Board Act, as codified at 49 U.S.C. Section 1154(b), precludes the admission into evidence or use of NTSB reports related to an incident or accident in a civil action for damages resulting from a matter mentioned in the report.

Contents

Figures

Acronyms and Abbreviations

AASHTO	American Association of State Highway and Transportation Officials
ABS	antilock brake system
ADT	average daily traffic
AOI	area of impact
AWE	Automated Waste Equipment
BASIC	behavioral analysis and safety improvement category [FMCSA]
BCE	Burlington County Engineers Office
BCR	Burlington County Road
BMI	body mass index
CAMI	Civil Aerospace Medical Institute
CAMP	Crash Avoidance Metrics Partnership [V2V–V2I]
CDL	commercial driver's license
CFR	*Code of Federal Regulations*
CMV	commercial motor vehicle
CSA	Compliance, Safety, Accountability [FMCSA program]
CTPD	Chesterfield Township Police Department
CVSA	Commercial Vehicle Safety Alliance
DAC	driver acceptance clinic [V2V–V2I]
DOT	US Department of Transportation
DSRC	dedicated short-range communication
ECM	electronic control module
EECU	electronic engine control unit
EMS	emergency medical service
EST	eastern standard time
FAA	Federal Aviation Administration
FARS	Fatality Analysis Reporting System
FHWA	Federal Highway Administration
FMCSA	Federal Motor Carrier Safety Administration

FMCSRs	*Federal Motor Carrier Safety Regulations*
FMVSS	*Federal Motor Vehicle Safety Standards*
FR	*Federal Register*
g	acceleration due to gravity
GAO	US Government Accountability Office
GAWR	gross axle weight rating
GCWR	gross combination weight rating
GES	General Estimates System
GHz	gigahertz
GPS	global positioning system
GST	Garden State Transport Corporation
GVWR	gross vehicle weight rating
IC	incident commander
mg	milligram
m-smac	Simulation Model of Automobile Collisions [formerly SMAC]
MCCRU	Motor Coach Compliance Review Unit [NJSP]
MD	model deployment [V2V–V2I]
MUTCD	*Manual on Uniform Traffic Control Devices*
NASDPTS	National Association of State Directors of Pupil Transportation Services
NASS	National Automotive Sampling System
NBCRSD	Northern Burlington County Regional School District
NHTSA	National Highway Traffic Safety Administration
NJDMV	New Jersey Department of Motor Vehicles
NJDOE	New Jersey Department of Education
NJDOT	New Jersey Department of Transportation
NJMVC	New Jersey Motor Vehicle Commission
NJSP	New Jersey State Police
NPRM	notice of proposed rulemaking
NTEA	National Truck Equipment Association
NTIA	National Telecommunications and Information Administration

NTSB	National Transportation Safety Board
OOS	out-of-service
OSA	obstructive sleep apnea
PARO	Palfinger American Roll-off
PCP	phencyclidine
psi	pounds per square inch
PTO	power takeoff
RDS	role delineation study [FMCSA]
SAE	SAE International
SBMTC	School Bus Manufacturers Technical Council
SI	sacroiliac [joint]
SNRI	serotonin and norepinephrine reuptake inhibitor
SSRI	selective serotonin reuptake inhibitor
U-NII	unlicensed national information infrastructure
U.S.C.	*United States Code*
USDOT	US Department of Transportation [number]
UTC	Coordinated Universal Time
V2V	vehicle-to-vehicle
V2I	vehicle-to-infrastructure
VECU	vehicle electronic control unit
VER	video event recorder

Executive Summary

On Thursday, February 16, 2012, about 8:15 a.m. eastern standard time, near Chesterfield, New Jersey, a Garden State Transport Corporation 2012 IC Bus, LLC, school bus was transporting 25 kindergarten–sixth-grade students to Chesterfield Elementary School. The bus was traveling north on Burlington County Road (BCR) 660 through the intersection with BCR 528, while a Herman's Trucking Inc. 2004 Mack roll-off[1] truck with a fully loaded dump container was traveling east on BCR 528, approaching the intersection. The school bus driver had stopped at the flashing red traffic beacon and STOP sign. As the bus pulled away from just forward of the white stop line on BCR 660 and entered the intersection, it failed to yield to the truck and was struck behind the left rear axle. The bus rotated nearly 180 degrees and subsequently struck a traffic beacon support pole. One bus passenger was killed. Five bus passengers sustained serious injuries, 10 bus passengers and the bus driver received minor injuries, and nine bus passengers and the truck driver were uninjured.

The National Transportation Safety Board (NTSB) determines that the probable cause of the Chesterfield, New Jersey, crash was the school bus driver's failure to observe the Mack roll-off truck, which was approaching the intersection within a hazardous proximity. Contributing to the school bus driver's reduced vigilance were cognitive decrements due to fatigue as a result of acute sleep loss, chronic sleep debt, and poor sleep quality, in combination with, and exacerbated by, sedative side effects from his use of prescription medications. Contributing to the severity of the crash was the truck driver's operation of his vehicle in excess of the posted speed limit, in addition to his failure to ensure that the weight of the vehicle was within allowable operating restrictions. Further contributing to the severity of the crash were the defective brakes on the truck and its overweight condition due to poor vehicle oversight by Herman's Trucking, along with improper installation of the lift axle brake system by the final stage manufacturer—all of which degraded the truck's braking performance. Contributing to the severity of passenger injuries were the nonuse or misuse of school bus passenger lap belts; the lack of passenger protection from interior sidewalls, sidewall components, and seat frames; and the high lateral and rotational forces in the back portion of the bus.

The crash investigation focused on the following safety issues:

- **School bus driver fatigue, sedating prescription medications, medical conditions, and commercial driver's license medical examinations:** These factors were examined to assess what might have caused the school bus driver to proceed into the intersection despite having adequate sight distance after stopping for the red traffic beacon and STOP sign.

---

[1] A *roll-off container* is typically an open steel receptacle used to remove and contain construction and demolition waste. Roll-off (also known as roll-off cable hoist) service refers to hydraulically operated rails and a cable hoist that are designed to load rectangular dumpsters atop the truck chassis behind the cab. Hydraulic pistons elevate and lower the forward end of the rails to work in concert with the cable hoist to raise or lower the dumpster onto the back of the truck.

- **Truck driver speed, oversight of overweight commercial vehicles, brake maintenance, and final stage manufacturing air brake system installation:** The final stage manufacturer improperly installed the lift axle air brakes on the truck, which—along with the condition of the brakes, the overloading of the vehicle, and the truck's speed—led to the severity of the collision with the school bus.

- **Connected vehicle technology:** Effective countermeasures are needed to assist in preventing intersection crashes—for example, systems such as connected vehicle technology could have provided an active warning to the school bus driver of the approaching truck as he began to cross the intersection. Although the bus driver was adamant in his postcrash interview that he had pulled forward sufficiently to see clearly in both directions, he failed to see the oncoming truck and proceeded into its path.

- **School bus occupant injuries and school bus crashworthiness:** The truck striking the school bus, as well as the bus striking the traffic beacon support pole, created high lateral forces that led to penetration of the bus interior. These factors contributed to the one fatality and severe injuries. Although the school bus was equipped with lap belts, the NTSB sought to determine how many passengers were using their seat belts, and examined whether properly worn lap belts and interior school bus protection measures could have improved the crashworthiness of the bus and mitigated passenger injury.

As a result of this crash investigation, the NTSB makes recommendations to the Federal Motor Carrier Safety Administration (FMCSA); National Highway Traffic Safety Administration (NHTSA); states of California, Florida, Louisiana, New Jersey, New York, and Texas; National Truck Equipment Association; National Association of State Directors of Pupil Transportation Services; National Association for Pupil Transportation; National School Transportation Association; School Bus Manufacturers Technical Council; National Safety Council, School Transportation Section; and Herman's Trucking Inc. The NTSB reiterates four recommendations to the FMCSA and three recommendations to NHTSA.

1 Factual Information

1.1 Crash Narrative

About 6:39 a.m.[1] on February 16, 2012, in Chesterfield, New Jersey, a 66-year-old school bus driver began transporting students to Northern Burlington County Regional High School on his regular morning route. He drove a 2012 IC Bus, LLC,[2] 54-passenger full-size school bus, which was owned by Garden State Transport Corporation (GST) and operated by Garden State Transport, Inc. After dropping off the students, he departed the high school and by 7:55 a.m. began his first scheduled pickup for Chesterfield Elementary School. The bus driver stated that this was his regularly scheduled route. He was on his 13th day of employment and had completed this route on each of his days on duty.[3] The crash occurred at 8:15 a.m., after the bus driver had picked up 25 students and as he was traveling on Burlington County Road (BCR) 660, crossing BCR 528, to complete his last three stops before heading to Chesterfield Elementary School.

The school bus driver was traveling north on BCR 660 when he approached the BCR 528–660 intersection, where BCR 660 was controlled by a STOP sign and a traffic control beacon with flashing red lights. Traffic on BCR 528 was controlled by the same traffic beacon with flashing yellow lights.[4] The bus driver stopped approximately 14 feet beyond the stop line, which was 16 feet from the edge of the Mack roll-off truck's travel lane on BCR 528.[5] He then proceeded into the intersection. At the same time, a fully loaded truck operated by Herman's Trucking Inc. was traveling east on BCR 528. The posted speed limit was 45 mph. (See figures 1 and 2.) The truck was occupied by its 38-year-old driver. As the school bus traveled through the intersection, the truck driver steered to the left and applied his brakes before striking the left rear section of the bus.

---

[1] Unless otherwise specified, all times in this report are eastern standard time (EST).

[2] IC Bus, LLC, is a division of Navistar, Inc.

[3] His first day of driving this route was Tuesday, January 31, 2012.

[4] An *intersection control beacon*—a traffic signal with one or more signal sections that operates in a flashing mode—is used only at an intersection to control two or more directions of travel.

[5] A *stop line* is a solid white pavement marking extending across approach lanes to indicate the point at which a stop is intended or required to be made.

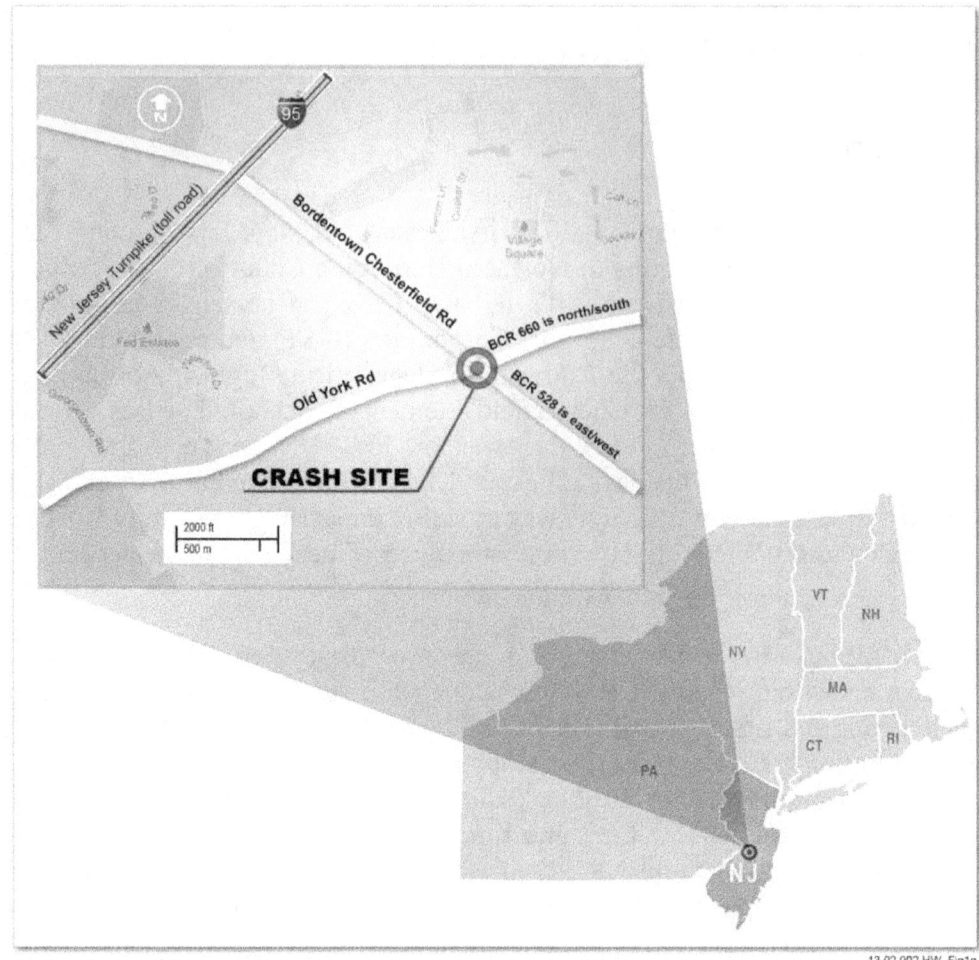

Figure 1. Regional map and view of crash site.

13.02.002.HW_Fig2a

Figure 2. Aerial view of BCR 528–660 intersection, with Bordentown–Chesterfield Road running east–west and Old York Road running north–south. (Google imagery, September 20, 2010)

The school bus, while rotating counterclockwise, continued moving another 50 feet to the northeast corner of the intersection, where its right side (behind the rear axle) struck a metal pole supporting a flashing intersection control beacon. The bus came to rest upright and facing west. The truck continued another 188 feet after impact and came to rest in a field northeast of the intersection, just off BCR 528.[6] (See figures 3 and 4.)

---

[6] This distance was 188 feet total, including 111 feet on the roadway and 77 feet off the pavement.

Figure 3. School bus at BCR 528–660 intersection on February 16, 2012. (Courtesy of Burlington County Prosecutor's Office)

Figure 4. School bus and truck (marked with green arrow) at final rest. (Courtesy of Burlington County Prosecutor's Office)

During the postcrash interview with investigating police and National Transportation Safety Board (NTSB) investigators, the school bus driver stated that nothing seemed out of the ordinary as he traveled north on BCR 660 and approached the intersection with BCR 528. He remembered the flashing red lights and the STOP sign. The bus driver could not recall if there was any crossing traffic as he approached the intersection, or if there was traffic on the opposite side of the intersection. During his interview, he moved his head to the left 90 degrees to demonstrate how he looked for traffic on BCR 528. He stated that he remembered "stopping where I felt safe. I looked left, looked right, looked left. I know where to stop so I can feel comfortable looking at the road. I stopped where I could see in both directions before the red flashing lights. Then I proceeded through the intersection." The driver also stated that once the road was clear, he wanted "to get across as quickly as possible before something is coming."

The bus driver continued northbound through the intersection. He stated that he then felt an impact to the rear driver side of the bus and the "bus went into the air," he looked into the rear mirror and observed the students "bouncing around," and finally he "felt the bus strike something else on the rear passenger side" before coming to rest.

The truck driver said that he was familiar with the BCR 528–660 intersection. In describing the crash, he stated that he was driving about 45 mph eastbound on BCR 528, approaching the intersection. He saw the school bus but could not tell whether the bus had just pulled up or whether it had made a complete stop because a line of trees obstructed his view of BCR 660 to the south. (See figure 5.) He said that he observed the bus as it entered the intersection and attempted to avoid the collision by steering to the left. He then applied the brakes and struck the school bus. (See figures 6 and 7.)

Figure 5. Truck driver view of BCR 528–660 intersection from 463 feet, showing exemplar school bus marked with yellow circle.

Figure 6. FARO Focus 3D laser scanner image of crash intersection facing westbound BCR 528, showing north and south lanes of BCR 660.

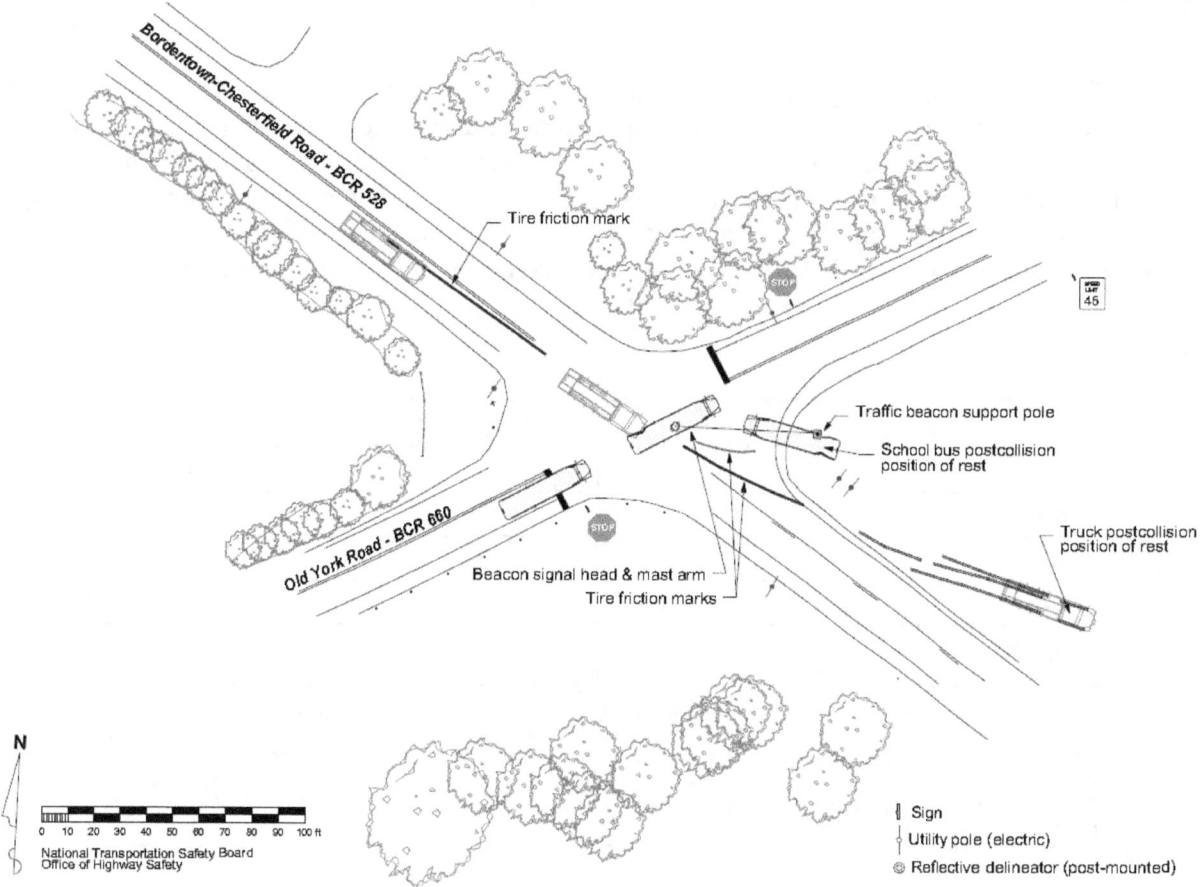

Figure 7. Crash scene diagram.

The crash scene, related environment, sight conditions, and both vehicles were examined by the NTSB and scanned using a FARO Focus 3D laser scanner. The scanner was placed on a tripod and automatically rotated 360 degrees to record its entire surrounding environment, emitting a laser beam to a range of about 300 feet. By keeping track of its position and orientation, and measuring the time it takes for the beam to reflect off objects, the scanner creates a three-dimensional view. Precise measurements can be obtained from the scans, such as the length of tire marks, deformations in damaged vehicles, and locations of site obstructions.

1.2 Injuries

One 11-year-old bus passenger was killed. Five bus passengers sustained serious injuries, 10 bus passengers and the bus driver received minor injuries, and nine bus passengers and the truck driver were uninjured. The fatally injured passenger sustained a posterior skull fracture, posterior scalp laceration, chin abrasion, and bilateral lower leg posterior abrasions.[7] Brain injuries, skull fractures, and thorax and extremity fractures accounted for the serious injuries. Minor injuries consisted of abrasions, contusions, lacerations, and strains. (See table 1.)

Table 1. Injuries.

Injury[a]	Drivers	Passengers	Total
Fatal	0	1	1
Serious	0	5	5
Minor	1	10	11
None	1	9	10
Total	2	25	27

[a]Title 49 *Code of Federal Regulations* (CFR) 830.2 defines fatal injury as any injury that results in death within 30 days of the accident. It defines serious injury as an injury that requires hospitalization for more than 48 hours, commencing within 7 days of the date of injury; results in a fracture of any bone (except simple fractures of fingers, toes, or nose); causes severe hemorrhages, or nerve, muscle, or tendon damage; involves any internal organ; or involves second- or third-degree burns, or any burn affecting more than 5 percent of the body surface.

Information on the 25 bus passenger seating positions was obtained from passenger interviews, witness statements, the Northern Burlington County Regional School District (NBCRSD) seating chart, and first responder interviews. The truck colliding with the school bus caused the first area of impact (AOI), and the bus striking the traffic beacon support pole caused the second AOI. (See figure 8.)

---

[7] Injury information was gathered from a noninvasive external examination; a full autopsy was not performed.

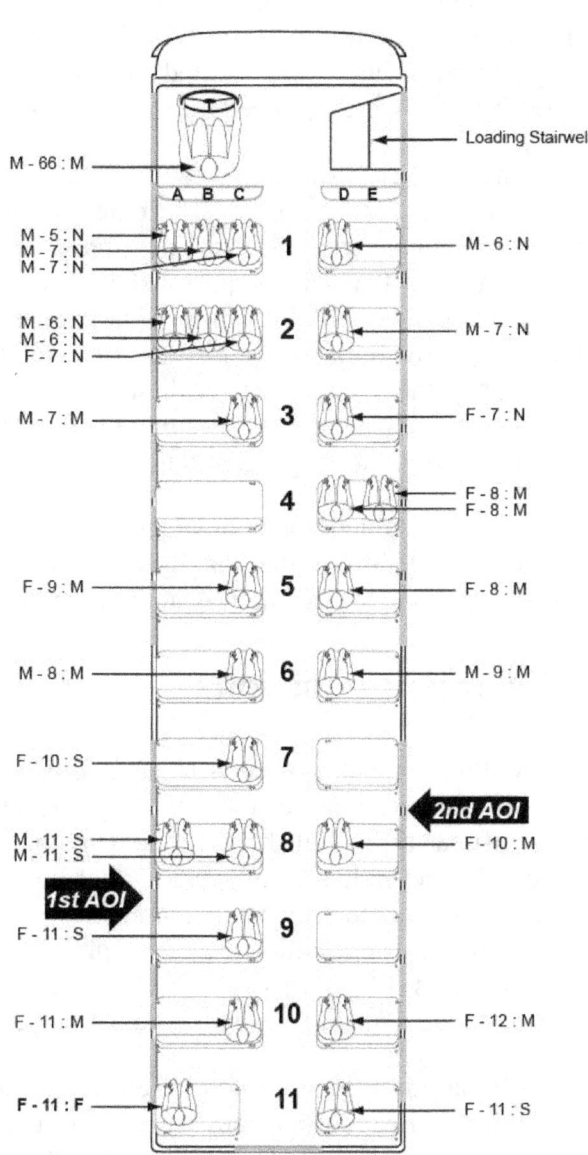

Figure 8. School bus passenger seating chart, injury, and demographic information.

Each school bus passenger seat was equipped with a lap belt (two-point restraint).[8] On the driver side, the bus featured 10 rows of passenger bench seats with three seating positions each and an 11th row with a two-position bench seat. Eleven rows of bench seats with two seating positions each were located on the passenger side. All 242 school buses owned and operated by GST were equipped with lap belts in each seating position.

The number of students wearing their seat belts at the time of the crash is not known. One witness and one first responder who entered the school bus immediately after the crash stated that they observed one child (a female, driver side, row 7) who "still had her seat belt on and she was hanging into the aisle because her belt was very loose"—and who was "lying in the aisle with her seat belt on still." Another witness observed a girl "lying on the floor with her seat belt still on" about five or six rows forward from the back, on the driver side. Other observers, along with a third witness, could not recall any other children wearing their seat belts; they noted that the children seated in the first rows remained in their seats, while those to the rear appeared to be lying across their seats or piled into the aisle.

The NTSB sent out questionnaires to the students on the bus; of the 21 responses received, 19 students reported that they had been wearing their seat belts at the time of the crash. One student responded that he was not wearing his belt.

1.3 Emergency Response

Within 2 minutes of the 8:15 a.m. collision, witnesses had notified the Burlington County central communications dispatcher. A Chesterfield Township Police Department (CTPD) off-duty officer arrived at the scene and radioed the dispatcher by 8:19 a.m. to also report the crash and that several students were unconscious. At 8:20 a.m., the dispatcher had contacted the CTPD, the Mansfield Township ambulance service, and the Crosswicks Fire Company. The CTPD chief and two additional officers arrived on scene by 8:21 a.m., and fire and emergency medical services (EMS) arrived at 8:22 a.m. The New Jersey State Police (NJSP) was notified of the crash at 8:53 a.m. and was on scene by 8:55 a.m.

Mutual aid was requested by the on-scene EMS incident commander (IC) by 8:29 a.m. An additional five ambulances were automatically dispatched, followed by another five ambulances dispatched by the EMS IC. In total, four law enforcement agencies, 14 ambulances

---

[8] Currently, six states require seat belts on buses. California requires lap-shoulder belts; and Florida, Louisiana, New Jersey, and New York require lap belt restraints. Texas requires lap-shoulder belts per *Texas Transportation Code Annotated*, section 547.701(e–f). Although the requirement applies to each bus purchased by a school district starting September 1, 2010, and chartered buses contracted for use by a school district starting September 2, 2011, a district is required to comply with this requirement only to the extent that the legislature has appropriate money to reimburse the district for expenses incurred. *New Jersey Statutes*, Motor Vehicles and Traffic Regulation, section 39:3B-10, states that students are required to wear the provided seat belts and that the school bus company and driver are not held liable when a child does not wear the seat belt. See also *New Jersey Statutes*, section 39:3B-11.

from 12 ambulance service companies, and two fire departments with three units responded to the crash.[9]

1.4 Crash Witnesses

There were several witnesses to the crash. One witness stated that she was located on BCR 660 southbound (facing the school bus), waiting to cross BCR 528. She stated that due to a blind spot at the intersection, she had to pull forward past the white stop line to see traffic on BCR 528. She looked to her right, observing the oncoming truck, though "it was not right on them," but she did not enter the intersection. She then faced forward and observed the bus beginning to slowly enter the intersection, and she thought that the bus was not going to make it. She then saw the truck strike the bus and immediately exited her vehicle to assist.

A second witness was located behind the school bus as it stopped at the intersection. This driver observed the bus driver stop and then pull forward. She did not notice the truck until just before it struck the bus. A third witness was an off-duty Burlington County Prosecutor's Office detective who was a few cars behind the bus on BCR 660. He observed the truck just as it entered the intersection and struck the bus. He then drove into the intersection, blocked traffic with his car, entered the bus to assess the passengers, and called 911.

1.5 Driver Information

1.5.1 Toxicology

Following this crash, blood samples were drawn from both drivers at the hospital. These samples were split with the NTSB and sent to the Federal Aviation Administration (FAA) Civil Aerospace Medical Institute (CAMI) for toxicological analysis. The truck driver's sample was determined to be negative for alcohol and all screened drug classes.[10] The school bus driver's sample, taken 3.75 hours postcrash, was determined to be negative for alcohol (no ethanol or major drugs of abuse) and positive for 7-amino-clonazepam, desmethylvenlafaxine (O-),[11] and tramadol.

Clonazepam is in a class of schedule I controlled substance medications called benzodiazepines and works by decreasing abnormal electrical activity in the brain; it is used to treat anxiety and seizure disorders. The predominant active metabolite of clonazepam is 7-amino-clonazepam. *Desvenlafaxine* is in a class of medications called serotonin and norepinephrine reuptake inhibitors (SNRI); it is also a metabolite of venlafaxine. It works by

---

[9] All 25 school bus passengers and the driver were transported to the hospital via ground ambulance. Two air medical evacuation helicopters responded; however, due to weather conditions that delayed their arrival, the passengers had already been transported from the scene.

[10] CAMI screened the drivers' blood samples for several drug classes, including amphetamines, opiates, marijuana, cocaine, phencyclidine (PCP), benzodiazepines, barbiturates, antidepressants, and antihistamines. For comprehensive information concerning all drugs detected by the laboratory, see the CAMI Drug Information website: jag.cami.jccbi.gov/toxicology/, accessed June 10, 2013.

[11] Desmethylvenlafaxine is a metabolite of venlafaxine.

increasing the amounts of serotonin and norepinephrine, natural substances in the brain that help maintain mental balance, and is used to treat depressive disorders.[12] *Tramadol* is in a class of medications called opiate agonists and works by changing the way the body senses pain.[13]

1.5.2 School Bus Driver

1.5.2.1 Certification, License, and Driving History. The 66-year-old school bus driver held a New Jersey class "B" commercial driver's license (CDL) with "P" passenger and "S" school bus endorsements and restrictions for corrective lenses and air brakes.[14] His medical examiner's certificate was issued on January 10, 2012, with an expiration date of January 10, 2013.[15] His first CDL was issued in January 2012, with an expiration date of January 2013. His driving record showed that he had received a traffic citation in March 2007 for "obstructing passage of another vehicle" and for a related crash on the same day. He had no suspensions or revocations.

1.5.2.2 Employment Background. The school bus driver had been employed with Garden State Transport, Inc., since January 30, 2012; the crash occurred on his 14th day of employment, the 13th day of driving on his own. Before his employment with the company, he was an ironworker from 1978 until his retirement in 2007. During his postcrash interview, he stated that he had been concerned about finances and had taken security guard and school bus driver positions to make extra money.[16] Company officials reported that he worked full-time and was paid by the hour. The driver was assigned a specific bus and a specific route for the school term. The driver's morning shift typically began when he arrived at the GST yard, conducted a pretrip bus inspection, and then left at 6:15 a.m. for his first scheduled pickup. The usual morning shift ended at 9:10 a.m. when he returned to the GST lot and left the bus. His usual afternoon shift began after 1:00 p.m. when he arrived at the GST lot to conduct a pretrip inspection prior to picking up students at the high school, dropping them at their designated stops, and then returning to Chesterfield Elementary to pick up students.[17]

---

[12] See www.nlm.nih.gov/medlineplus/druginfo/meds/a608022 html, accessed April 4, 2013.

[13] See www.nlm.nih.gov/medlineplus/druginfo/meds/a695011 html, accessed April 4, 2013.

[14] The New Jersey class B CDL allows the operation of any vehicle with a gross vehicle weight rating (GVWR) of 26,001 or more pounds; a vehicle with a GVWR of 26,001 pounds or more towing a trailer with a GVWR of 10,000 pounds or less; or a bus with a GVWR of 26,001 pounds or more designed to transport 16 or more passengers. A class B CDL also allows the operation of class C vehicles with the proper endorsements. The bus driver's corrected vision was 20/30 in both his left and right eyes, measured using a Snellen visual acuity test. In a statement to investigators, the driver reported that he had been wearing his glasses while driving. The air brake restriction means that the driver cannot operate a vehicle equipped with air brakes; the school bus was equipped with hydraulic brakes.

[15] The *Federal Motor Carrier Safety Regulations* (FMCSRs), at 49 CFR 391.41, require that commercial drivers be medically certified as physically qualified to drive. The school bus driver had been certified by a doctor of chiropractic, but—though he was found to meet the medical certification standards—due to hypertension, he required yearly evaluation rather than the otherwise allowable biennial evaluation.

[16] The bus driver worked briefly as a security guard from September 2011 until starting with Garden State Transport, Inc.

[17] The driver's route included 32 high school bus stops, excluding Northern Burlington County Regional High School, and 25 elementary school bus stops, excluding Chesterfield Elementary School.

1.5.2.3 Precrash Activities. On the Monday, Tuesday, and Wednesday preceding the crash, the school bus driver drove his usual route. See table 2 and figure 9 for a summary of the driver's precrash activities.

Table 2. School bus driver's precrash activities (February 13–16, 2012).

Monday, February 13, 2012		
Time (EST)	**Activities**	**Source**
4:30 a.m.	Awakens by alarm clock	Interview
5:30 a.m.	Departs home for GST bus lot	Interview
Unknown	Conducts pretrip bus inspection	Interview
6:15 a.m.	Departs GST bus lot for first route	Interview
6:39 a.m.	First high school pickup	Schedule
7:12 a.m.	Last high school pickup	Schedule
7:16 a.m.	Drops students at high school	Schedule
7:29 a.m.	Makes outgoing call (first call of day)	Cell phone records
7:50 a.m.	Departs high school	Schedule
7:55 a.m.	First elementary school pickup	Schedule
8:12 a.m.	Last elementary school pickup	Schedule
8:15 a.m.	*Scheduled* student dropoff at elementary school	Schedule
8:20 a.m.	*Actually arrives* at elementary school	Interview
8:25 a.m.	Departs elementary school	Schedule
9:10 a.m.	Arrives at GST bus lot	Interview
Unknown	Returns home for lunch	Interview
Unknown	Naps 10–15 minutes	Interview
~12:40 p.m.	Departs home for GST bus lot	Interview
1:00 p.m.	Arrives at GST bus lot	Interview
Unknown	Completes afternoon pretrip bus inspection	Interview
1:45 p.m.	Arrives at high school	Interview
~2:15 p.m.	High school students loading	Interview
2:20 p.m.	Departs high school	Interview
2:45 p.m.	Last scheduled high school dropoff	Interview
3:05 p.m.	Arrives at elementary school	Interview
3:30 p.m.	Departs elementary school	Interview
4:00 p.m.	Arrives at GST bus lot	Interview
Unknown	Performs posttrip inspection, turns in paperwork and bus keys	Interview
4:20 p.m.	Departs GST to return home	Interview
5:00 p.m.	Arrives at home	Interview
7:12 p.m.	Receives incoming call (last call of day)	Cell phone records
~11:15 p.m.	Goes to bed	Interview
Unknown	Wakes up once or twice to use bathroom	Interview
Tuesday, February 14, 2012		
Time (EST)	**Activities**	**Source**
4:30 a.m.–4:00 p.m.	Maintains same schedule as Monday	Interview, schedule
Unknown	Performs posttrip inspection, turns in paperwork and bus keys	Interview
4:20 p.m.	Departs GST to return home	Interview
Unknown	Arrives at home	Interview
Unknown	Visits brother-in-law at hospital	Interview
~11:15 p.m.	Goes to bed	Interview
Unknown	Wakes up once or twice to use bathroom	Interview

Wednesday, February 15, 2012		
Time (EST)	**Activities**	**Source**
4:30 a.m.	Awakens by alarm clock	Interview
5:30 a.m.	Departs home for GST bus lot	Interview
Unknown	Returns home to retrieve item, departs	Interview
Unknown	Conducts pretrip bus inspection	Interview
6:15 a.m.	Departs GST bus lot for first route	Interview
6:39 a.m.–9:10 a.m.	Maintains same schedule as Monday, Tuesday	Interview, schedule
9:59 a.m.	Calls voicemail (first call of day)	Cell phone records
Unknown	Returns home for lunch	Interview
Unknown	Naps 10–15 minutes	Interview
~12:40 p.m.	Departs home for GST bus lot	Interview
1:00 p.m.	Arrives at GST bus lot	Interview
1:13 p.m.	Calls voicemail (last call of day)	Cell phone records
1:45 p.m.–4:20 p.m.	Maintains same schedule as Monday, Tuesday	Interview, schedule
Unknown	Arrives at home	Interview
~11:15 p.m.	Goes to bed	Interview
Unknown	Wakes up once or twice to use bathroom	Interview
Thursday, February 16, 2012		
Time (EST)	**Activities**	**Source**
4:30 a.m.	Awakens by alarm clock	Interview
5:30 a.m.	Departs home for GST bus lot	Interview
6:13 a.m.	Makes outgoing call (first call of day)	Cell phone records
6:15 a.m.	Departs GST bus lot for first route	Interview
6:39 a.m.	*Scheduled* high school pickup	Schedule
6:40 a.m.	*Actually arrives* for first high school pickup	Interview
7:12 a.m.	Last high school pickup	Schedule
7:15 a.m.	Arrives at high school	Interview
7:16 a.m.	Drops students at high school	Schedule
7:50 a.m.	Departs high school	Schedule
7:55 a.m.	First elementary school pickup	Schedule
8:12 a.m.	Last elementary school pickup	Schedule
8:15 a.m.	*Scheduled* student dropoff at elementary school	Schedule
8:15 a.m.	**Crash at BCR 528–660 intersection**	

Figure 9 provides a graphic activity history for the three days preceding the crash and is based on information obtained during the school bus driver's interview, from his work schedule, and from his cell phone records. The bus driver kept a fairly regular sleep schedule, retiring about 11:15 p.m. and awaking at 4:30 a.m. on each of the three days prior to the crash, which provides about 5.25 hours of time in bed. On his days off, he would usually go to bed at approximately the same time but awake about 6:30 a.m. When asked about napping, the bus driver told investigators that he took a 10–15 minute nap every day at lunch, which he found to be refreshing. He stated that he has never been diagnosed with sleep problems, including insomnia, obstructive sleep apnea (OSA), or restless legs syndrome. During the postcrash interview, the bus driver was upset over the loss of his brother-in-law, who had died the Tuesday before the crash.

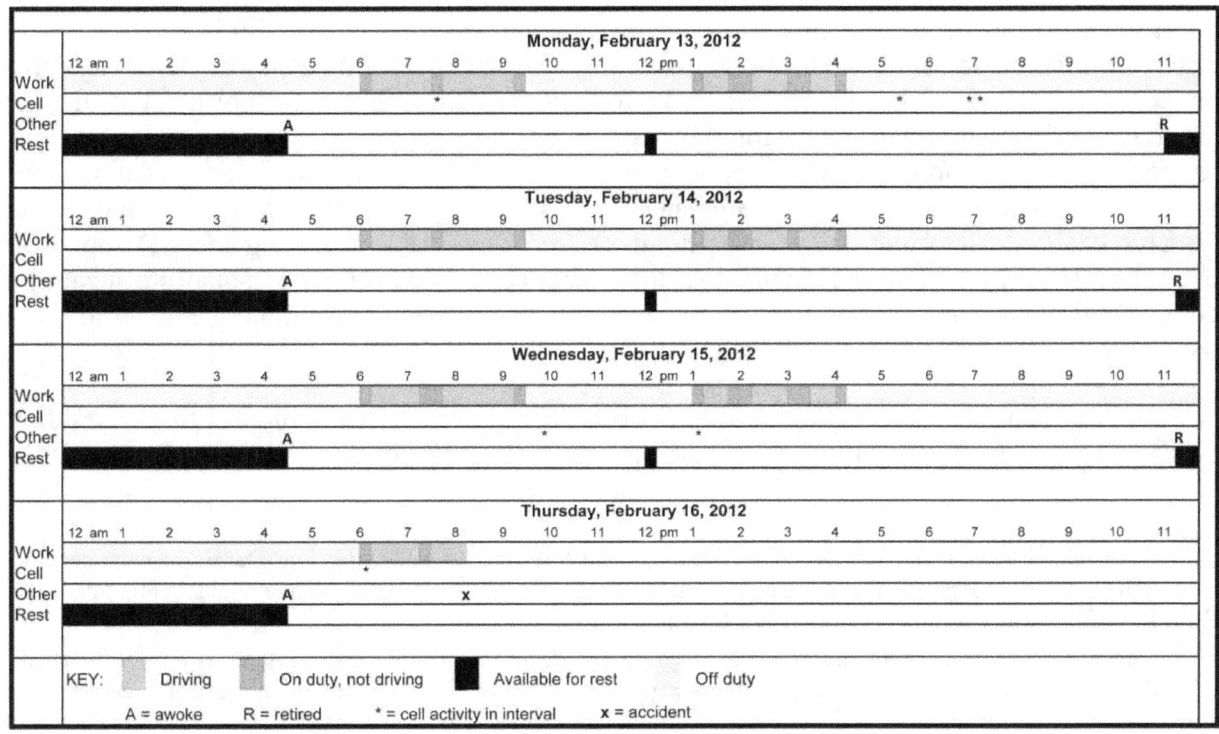

Figure 9. Summary of school bus driver's precrash duty schedule.

1.5.2.4 Medical History. When interviewed postcrash, the school bus driver stated that he did not go to his personal physician for the CDL medical examination. On January 10, 2012, he went to a doctor of chiropractic to whom he was referred by another GST driver, explaining that because the chiropractor had all the forms, it would be easier and faster than seeing his personal physician.

The school bus driver checked "yes" on his medical certificate form for the following listed conditions: illness or injury in the last five years; heart disease or heart attack or other cardiovascular condition (indicated that he was taking simvastatin); heart surgery, valve replacement/bypass, angioplasty, or pacemaker (driver crossed through "replacement" and wrote "repair"); digestive problems; nervous or psychiatric disorders, for example, severe depression (he crossed through "severe" and wrote "mild" and indicated that he was taking clonazepam for anxiety); spinal injury or disease; and regular, frequent alcohol use.[18] He also listed citalopram but did not indicate a medical condition. The NTSB obtained copies of the driver's current prescriptions and their prescribed dosage (in milligrams [mg]). The driver was prescribed 40 mg simvastatin daily, 50 mg metoprolol ER daily, 4 mg esomeprazole twice daily, 30 mg lansoprazole twice daily, 1 mg clonazepam twice daily, 20 mg citalopram daily, 50 mg desvenlafaxine daily, 50 mg tramadol four times daily, and 5 mg oxycodone/325 mg acetaminophen four times daily.[19] The CDL medical examiner found that the driver met the standards as set forth in 49 CFR 391.41 but required periodic monitoring for high blood pressure (yearly) and a restriction for corrective lenses.

The CDL medical form stated that for any "yes" answer, the applicant was to indicate onset date, diagnosis, treating physician's name and address, and current limitations. There is also a requirement to list all medications (including over-the counter medications) used regularly or recently. The school bus driver checked "yes" to frequent, regular alcohol use but did not provide additional information, including his doctor's diagnosis of alcoholism three months prior to the CDL examination, nor did he report that he was warned of the risk of oversedation from mixing benzodiazepines with alcohol and not to take them at the same time.[20] On January 30, 2012, the same day he was hired by GST—20 days following the CDL medical exam and 16 days prior to the crash—the driver was again seen by his personal physician; he complained of anxiety, and the doctor's notes stated "switch" to desvenlafaxine at 50 mg and recheck in one month.

The school bus driver checked "no" on the CDL medical examination form regarding "chronic low back pain" and did not report his prescription use of tramadol (50 mg four times

---

[18] The school bus driver indicated on his CDL medical form that he consumed two glasses of wine per day. Postcrash, he stated to police that he had an estimated two "double scotches" on the night before the crash (February 15), between the hours of 7:30 and 11:30 p m.

[19] See section 1.5.1 for a brief discussion of clonazepam, desvenlafaxine, and tramadol. *Simvastatin* is in a class of medications called HMG-CoA reductase inhibitors (statins). It slows the production of cholesterol to decrease the amount that may build up on the walls of the arteries and block blood flow. *Metoprolol* is in a class of medications called beta blockers, which relax blood vessels and slow heart rate to improve blood flow and decrease blood pressure. *Esomeprazole* is a prescription gastric parietal cell proton pump inhibitor used for the treatment of gastroesophageal reflux disease and other gastric ailments. *Lansoprazole* is a prescription proton pump inhibitor used for the treatment of gastroesophageal reflux disease or ulcers in those taking nonsteroidal anti-inflammatories, for the treatment of excess stomach acid, or for the treatment of ulcers caused by a particular bacterium. *Citalopram* is in a class of antidepressants called selective serotonin reuptake inhibitors (SSRI). It is thought to work by increasing the amount of serotonin, a natural substance in the brain that helps maintain mental balance. *Oxycodone/APAP* is a prescription combination of oxycodone (an opiate [narcotic] analgesic) and acetaminophen used to relieve moderate-to-severe pain. See www.ncbi.nlm.nih.gov/pubmedhealth/, accessed November 1, 2012.

[20] The school bus driver's medical records showed that on October 25, 2011, he was given a diagnosis of alcoholism, to be treated with clonazepam for withdrawal symptoms and citalopram for depression and anxiety. On November 3, 2011, the doctor indicated in the medical record that the driver was still experiencing depression and anxiety and would continue on the clonazepam and citalopram. The doctor indicated alcohol abuse in the record and encouraged the driver to abstain from alcohol and to seek therapy. On December 1, 2011, the doctor documented the patient's alcohol abuse, instructed him to continue on the two drugs, warned him of the risk of oversedation, and again recommended seeing a therapist.

daily) or oxycodone (5 mg with 325 mg acetaminophen four times daily) for low back and leg pain. On December 30, 2011, the driver had been seen by his orthopedic doctor, whose medical records show that he was experiencing pain along the sacroiliac (SI) joint and pain radiating down his right leg, and he was ambulating with progressive difficulty and a significant antalgic gait.[21] The doctor recommended that he be evaluated for possible lumbar spine epidural injections and possible spinal surgery.

During the postcrash interview with local law enforcement and the NTSB, the school bus driver reported that he took medication daily for high cholesterol, to relieve anxiety/panic attacks, to relieve pain, to treat depression, and to treat hypertension. He stated that though some of his medications warned against driving, they did not cause any side effects and he had been on the medications for enough time that he was accustomed to them. The driver reported taking two anxiety medications in the morning and one in the afternoon as soon as he got home from work. He took the rest of his medications at night. He reported that he did take his prescribed anxiety medications on the morning of the crash. According to the driver, his prescribing physicians did not place any restrictions on driving. The driver also stated that he had an enlarged prostate, which caused him to wake up during the night to use the bathroom.

1.5.2.5 GST Driver Training. Once hired, GST drivers receive carrier training on operation of the school bus they will be driving and on all other buses in the fleet that they may potentially drive. The school bus driver stated to NTSB investigators that he received training on operation of the bus and emergency evacuation, and that when he was proficient he took the New Jersey Department of Motor Vehicles (NJDMV) CDL road test in a GST-supplied school bus. GST told NTSB investigators that it had given the driver emergency evacuation training but could not provide any documentation of such.

The school bus driver then received hands-on training and rode along with an experienced GST driver for one day on a different route. On the second day of hands-on training, he operated the bus while the experienced driver observed him. (Drivers are required to drive their assigned route in the bus they will be using on that route prior to the school year and before picking up and transporting students.) The driver was then permitted to drive on his own. On behalf of the NBCRSD, GST provided the travel route, addresses of the required bus stops, times of pickup and dropoff of students, and a seating chart for the elementary school students.

GST's standard operating procedures indicate that "the training curriculum will be provided to the extent to enable each employee to reach the professional level of the specific job requirements." The school bus driver was given an employee handbook that included standard operating procedures, such as the prohibition on the use of cell phones while driving a school bus and the requirement to reference the *Regional District Driver's Handbook* for rules pertaining to a specific route. The GST driver handbook stated only the following regarding driver training:

> All driver and escort training will be received from a qualified instructor who will train, evaluate and record each employee's attendance and progress. The training curriculum will be provided to the extent required to enable each employee to

---

[21] An antalgic gait is a limp in which a phase of the gait is shortened on the injured side to alleviate the pain experienced when bearing weight on that side.

reach the professional level of their specific job requirements. All ongoing training and monthly training schedules will be posted on the bulletin board.

GST reported that it provides in-service training/safety meetings twice a year. The school bus driver had not yet received in-service training. The training material consisted of discussions and videos supplied by GST's insurance company. Topics included special needs, "no child left on bus," equipment on vehicle, and defensive driving.[22] GST drivers did participate in the NBCRSD twice-yearly school bus evacuation drills, which all school bus drivers are required to attend.[23]

1.5.3 Truck Driver

1.5.3.1 Certification, License, Driving History, and Medical History. The 38-year-old truck driver held a valid class A New Jersey CDL.[24] The license had no endorsements and was not subject to any restrictions. It was issued in December 2010 and expired in December 2014.[25] The truck driver's New Jersey Motor Vehicle Commission (NJMVC)-certified driver abstract (driving history) showed two crashes in 1991; a speeding violation in 1997; a reckless driving violation in July 2003; completion of a defensive driving course on November 1, 2003; a reckless driving violation with 90-day license suspension on November 13, 2003 (his license privilege was restored on February 2, 2004); and a crash in September 2006.[26,27]

The truck driver held a valid medical certification card issued in March 2010 by an internal medicine doctor and was qualified for 2 years. The driver stated to NTSB investigators that he did not have any medical or health conditions. He said that he did not take any prescription drugs, herbal supplements, or over-the-counter medications—both in general and specifically on the day of the crash. Pharmacy records did not indicate any current prescriptions. He responded "no" to all of the health conditions listed under the health history section of the medical certification form for his CDL fitness examination. The physician performing the exam indicated that he observed no abnormalities in any of the driver's general body systems.[28]

---

[22] The publication dates for these videos range from 1989 to 2003.

[23] School bus evacuation drills are held at a school when the bus drops off students. The students are directed by the school bus driver, assisted by school staff, to exit the vehicle by opening the emergency exit doors. Older students open the doors and assist younger students out of the bus. Emergency window and overhead emergency hatches are discussed but not operated. See *New Jersey Administrative Code* 6A:27-11.2, Student Transportation, evacuation drills and safety education.

[24] The New Jersey class A CDL allows the operation of a tractor trailer or a truck and trailer with a gross combination weight rating (GCWR) of 26,001 pounds or more, provided that the GCWR of the vehicle being towed is more than 10,000 pounds. A class A CDL also allows the operation of class B and C vehicles, with the proper endorsements.

[25] The truck driver stated to NTSB investigators that he had held a CDL for approximately 10 years.

[26] None of the violations occurred in a CMV.

[27] Herman's Trucking had reviewed the driver abstract from the NJDMV in 2010, which provided a five-year history. The abstract showed only the driver's 2006 crash and no traffic violations.

[28] General body systems are: general appearance; eyes, ears, mouth, and throat; heart, lungs, and chest; abdomen and viscera; vascular system; genitourinary; extremities; spine and other musculoskeletal; and neurological.

1.5.3.2 Employment Background. The driver of the truck had first been employed by Herman's Trucking in 2002 as a laborer and then as a truck driver. He did not attend a CDL driving school; he received his CDL permit and was then trained by Herman's Trucking. He left the company in 2007 and worked for a different trucking company until 2009, when he reapplied for a driving position. At the time of the crash, the driver was hauling construction waste to the company yard, a route he was very familiar with and had driven many times before.

1.5.3.3 Precrash Activities. Table 3 summarizes the truck driver's activities on the Monday, Tuesday, and Wednesday preceding the crash.[29] When interviewed postcrash, the truck driver described his quality of sleep as 7 or 8 on a scale of 1–10. His time of awakening varies, and he sleeps until 7:30 or 8:00 a.m. on Saturday and Sunday. He occasionally wakes at night to get a drink or something to eat but does not have any difficulty falling back to sleep. He has a three-month-old child but stated that the baby does not disturb his rest. Based on the truck driver's statements, he received 8–8.5 hours of sleep on the four nights prior to the crash.

---

[29] This recent driver activity history is based on an interview with the driver and his cell phone records.

Table 3. Truck driver's precrash activities (February 13–16, 2012).

Monday, February 13, 2012		
Time (EST)	**Activities**	**Source**
~6:00 a.m.	Awoken by his fiancee	Interview
6:04 a.m.	Sends text message (first of day)	Cell phone records
~6:30 a.m.	Departs home for work	Interview
10:56 a.m.	Makes outgoing call (first of day)	Cell phone records
3:30 p.m.	Ends workday	Interview
7:06 p.m.	Sends text message (last cell phone use of day)	Cell phone records
~9:30 p.m.	Goes to bed	Interview
Tuesday, February 14, 2012		
Time (EST)	**Activities**	**Source**
~5:45 a.m.	Awoken by his fiancee	Interview
5:54 a.m.	Sends text message (first of day)	Cell phone records
~6:30 a.m.	Departs home for work	Interview
3:51 p.m.	Receives phone call (last of day)	Cell phone records
~4:00 p.m.	Ends workday	Interview
~9:30 p.m.	Goes to bed	Interview
Wednesday, February 15, 2012		
Time (EST)	**Activities**	**Source**
5:30 a.m.	Awoken by his fiancee	Interview
~6:30 a.m.	Departs home for work	Interview
~4:00 p.m.	Ends workday	Interview
8:49 p.m.	Sends text message (last cell phone use of day)	Cell phone records
~9:30 p.m.	Goes to bed	Interview
Thursday, February 16, 2012		
Time (EST)	**Activities**	**Source**
~1:30 a.m.	Awakens for a drink of water	Interview
6:00 a.m.	Awoken by his fiancee	Interview
6:30 a.m.	Departs home for work	Interview
~6:40 a.m.	Arrives at work	Interview
6:48 a.m.	Makes outgoing call (first of day)	Cell phone records
Unknown	Completes pretrip vehicle inspection	Interview
Unknown	Departs company yard	Interview
Unknown	Arrives at construction site	Interview
~7:10 a.m.	Departs construction site with full container	Interview
~7:30 a.m.	Arrives at company yard	Inteview
~7:40 a.m.	Departs company yard	Interview
7:58 a.m.	Receives phone call, ends call about 8:07 a.m.	Cell phone records
~8:00 a.m.	Arrives at construction site	Interview
~8:10 a.m.	Departs construction site with full container	Interview
8:15 a.m.	**Crash at BCR 528–660 intersection**	

1.6 Postcrash Truck Vehicle Inspection

1.6.1 General Information and Damage

The 2004 four-axle Mack Trucks, Inc., Granite roll-off truck with snowplow mount was purchased by Herman's Trucking of Wrightstown, New Jersey, in February 2004.[30] The vehicle was manufactured in two stages: first, as a chassis only (considered an incomplete vehicle by Mack Trucks), and then with an additional axle (referred to as the second axle on this vehicle) and the roll-off bin container system. Automated Waste Equipment (AWE) was the final stage manufacturer. In 2009, Palfinger North America purchased AWE, which also worked under the trade name American–Roll-off. The company is now called Palfinger American Roll-off (also known as "PARO").[31] The truck was registered by Herman's Trucking with the state of New Jersey with four axles and a registered allowable maximum gross vehicle weight of 80,000 pounds.[32] The truck weighed 84,950 pounds postcrash, and the odometer reading was 246,843 miles.

Damage to the truck as a result of the collision consisted of deformation to the front steel bumper (rearward crushing deformation), the grill and radiator, the red fiberglass hood, the right headlight, and the snowplow mount attachment. Maximum deformation was measured at 10 inches rearward, and the damage spanned a 77-inch width across the front. (See figure 10.) Contact damage was observed on the snowplow mount, along with "school bus yellow" paint transfer. The right upright of the snowplow mount was displaced rearward a greater distance than the left upright, causing a twisting rearward displacement, which resulted in an induced rearward crush to the center portion of the steel bumper. The outer edges of the vertical uprights of the snowplow mount are approximately 25 inches apart, and the rear mounts for the lifting plate are located approximately 48 inches above ground level.

---

[30] At the time of the crash, the truck was equipped with a snowplow mounting bracket only—the snowplow was not attached. Under a snow removal services contract between Herman's Trucking and the state of New Jersey, vehicles designated for snow removal must have the snowplow mount attached from November 1 through April 30. The truck was one of the designated snow removal vehicles.

[31] See www.palfinger.com/usa/media/North%20America/Documents/world%20inserts/palfingerna_World-20_English.pdf?as=l&la=en-US, accessed June 11, 2013.

[32] The vehicle was manufactured by Mack Trucks with three axles and a manufacturer certification that it met the *Federal Motor Vehicle Safety Standards* (FMVSS). Manufacturers certify that their vehicle axle weights match the design capacity, and all states have laws prohibiting the loading of a vehicle in excess of its design capacity. The chassis was certified to have gross axle weight ratings (GAWR) of 18,000 pounds front axle, 20,000 pounds second axle, 24,680 pounds third axle, and 24,680 pounds rear tandem axle. The second axle, installed by AWE, could be retracted vertically when the vehicle was empty and lowered to increase the load-carrying capacity.

Figure 10. Front view of damaged truck.

The truck was not equipped with an event data recorder. It was equipped with an electronic engine control unit (EECU) and a vehicle electronic control unit (VECU).[33] Postcrash, both the EECU and the VECU were removed and downloaded by NTSB investigators. Data recovered from the modules were determined to be unrelated to the crash. The driver had been using a personal portable global positioning system (GPS) unit, which is capable of storing a detailed tracklog whenever the receiver has a lock on the GPS navigational signal. According to downloaded data, the truck driver had crossed the BCR 528–660 intersection three previous times on February 16, 2012: at 07:05 hours, 07:21 hours, and 07:57 hours. The calculated average speeds of the truck while crossing the intersection were 49.0 mph, 45.9 mph, and 48.1 mph, respectively. The posted speed limit was 45 mph. Just prior to the truck striking the school bus, the GPS unit indicated that the truck's average speed just before and while entering the intersection was 47.3 mph at 08:15:16 hours to 50.2 mph at 08:15:30 hours.[34] The GPS

---

[33] The purpose of these modules is to control engine timing and fuel injection based on engine and sensor inputs. The modules are also capable of diagnostics associated with engine or sensor faults, which may then illuminate warnings on the dash.

[34] At 07:05 hours, within 2,000 feet before and after passing through the intersection, the average speed of the truck was calculated at 46–51 mph; during the second pass, between 1,500–2,000 feet before and after the intersection, the average speed was calculated at 46–51 mph; and during the third pass, between 1,600–1,800 feet before and after the intersection, the average speed was calculated at 48–52 mph. The last recorded truck position before any of the collision events occurred was at 08:15:30 hours; during the fourth and final pass approach, at approximately 170 feet west of the intersection, the average speed preceding this point was 50.2 mph.

indicated that the average speed of the truck as it was in the intersection, struck the school bus, and departed the intersection was 50.2 mph at 08:15:30 hours to 47.0 mph at 08:15:35 hours and 40.8 mph at 08:15:36 hours.[35] (See appendix B.)

1.6.2 Mechanical Condition

All major mechanical systems were examined, including the steering, suspension, and braking systems. No damage was noted to any of the steering system components; all but one of the connections were solid and free of wear.[36] Herman's Trucking vehicles were required per 49 CFR 396.17 to be inspected annually, and records examined by NTSB postcrash revealed that these requirements were met.[37]

All but one tire had adequate tread depth in accordance with 49 CFR 393.75 specifications. Four of the 12 tires on the truck were underinflated but not to the point that the truck would be considered out of service using Commercial Vehicle Safety Alliance (CVSA) criteria.[38]

1.6.3 Vehicle Braking and Air Systems

The truck braking system was examined postcrash. Pushrod stroke measurements were taken for all of the brake chambers, and several brake-related defects were found. The pushrod stroke on the left side of axle 4 was over the adjustment limit by 0.25 inch, resulting in this brake being classified as defective according to CVSA OOS criteria. Herman's Trucking stated to NTSB investigators that its vehicles were inspected biweekly, and manual brake slack adjustments were performed if necessary. If an out-of-adjustment brake was equipped with an automatic slack adjuster, mechanics would adjust the slack adjuster one time and mark the vehicle frame with a grease pencil to indicate such. If that brake was found to be out of adjustment a second time, the slack adjuster would be replaced. The NTSB postcrash inspection also found a void in the edge of the brake pad lining on the left side of axle 2, which would be considered a defective lining condition—as would the loose brake pad lining on the right side of axle 3 and the brake components on the right side of axle 4, which were contaminated with oil or grease. These defective lining conditions would result in each brake being classified as defective under CVSA OOS criteria.

---

[35] Times are referenced to Coordinated Universal Time (UTC).

[36] The steering shaft was connected to the steering gear through a 45 degree transfer joint. The universal joint connecting the steering shaft to the 45 degree transfer joint showed some signs of wear. All other steering system components were free of wear and excess play.

[37] The annual US Department of Transportation (DOT) inspections were conducted in-house, by the head of the maintenance department. The vehicle inspection report dated February 25, 2009, indicated that the service brakes needed to be repaired. The report also indicated that those repairs were completed on the same date as the inspection. No defects were noted on either the February 25, 2010, or February 16, 2011, inspection reports. The truck was due to have its next annual vehicle inspection by the end of February 2012, the month in which this crash occurred.

[38] Only one tire did not meet the minimum tread depth requirement; therefore, the vehicle did not receive an out-of-service (OOS) order for tire tread depth. CVSA criteria do not consider a tire to be out of service until it has 50 percent less than the maximum inflation pressure listed on the tire sidewall.

As part of its air system, the truck had a pneumatic system that operated the brakes as well as the lift axle. The original air brake system installed on the chassis by Mack was later modified when the final stage manufacturer (AWE) installed the roll-off cable hoist system and lift axle. Modifications to the vehicle's braking system included plumbing the rear service brake relay valve so that it activated six brake chambers (four type 30 chambers and two type 24 chambers) rather than the four brake chambers originally supplied on the chassis by the vehicle manufacturer. A quick-release valve was also added to the braking system on the lift axle to assist in releasing air from the lift axle brake chambers. Other modifications to the air system included an additional reservoir tank (an expansion reservoir),[39] which was tied into the primary air system to increase the available air volume.[40] Air cushions responsible for lowering the lift axle were tied into the primary air system, which also supplied air to the brakes on the lift axle and both drive axles of the truck.

The final stage manufacturer (AWE) had installed the expansion air reservoir when the air-operated lift axle and associated air suspension and brake components were installed. This air reservoir served two functions: (1) because there was no "one-way" or single check valve between the primary and auxiliary air reservoirs, it increased the effective reservoir volume of the primary air system; and (2) it supplied air for the lift axle and the power takeoff (PTO) system, which was used to operate the roll-off hoist and winch for loading and unloading a container.[41] Air lines connected the primary air system reservoir directly to the expansion air reservoir, and a pressure protection valve separated the expansion air reservoir from the lift axle and PTO air supply lines. Figure 11 is a schematic of the layout of the truck's air system as well as its braking and antilock brake system (ABS) components.

---

[39] The term "expansion reservoir" is used in the *Installer's Guide for Mack Class 8 Chassis* and is used in this report for consistency.

[40] Other major air system components included a governor, air dryer, supply reservoir or "wet tank," primary air system reservoir, and secondary air system reservoir.

[41] The PTO system is ultimately powered by the engine through the transmission; however, the mechanical connection between the engine and the PTO is engaged by air from the expansion reservoir through control valves in the cab. When in use, the lift axle is lowered to the ground by air pressure in the air spring/suspension cushions, which overcomes the force of mechanical springs located on the axle assembly (used to raise the axle). When the lift axle is not in use, the mechanical springs return it to a raised position.

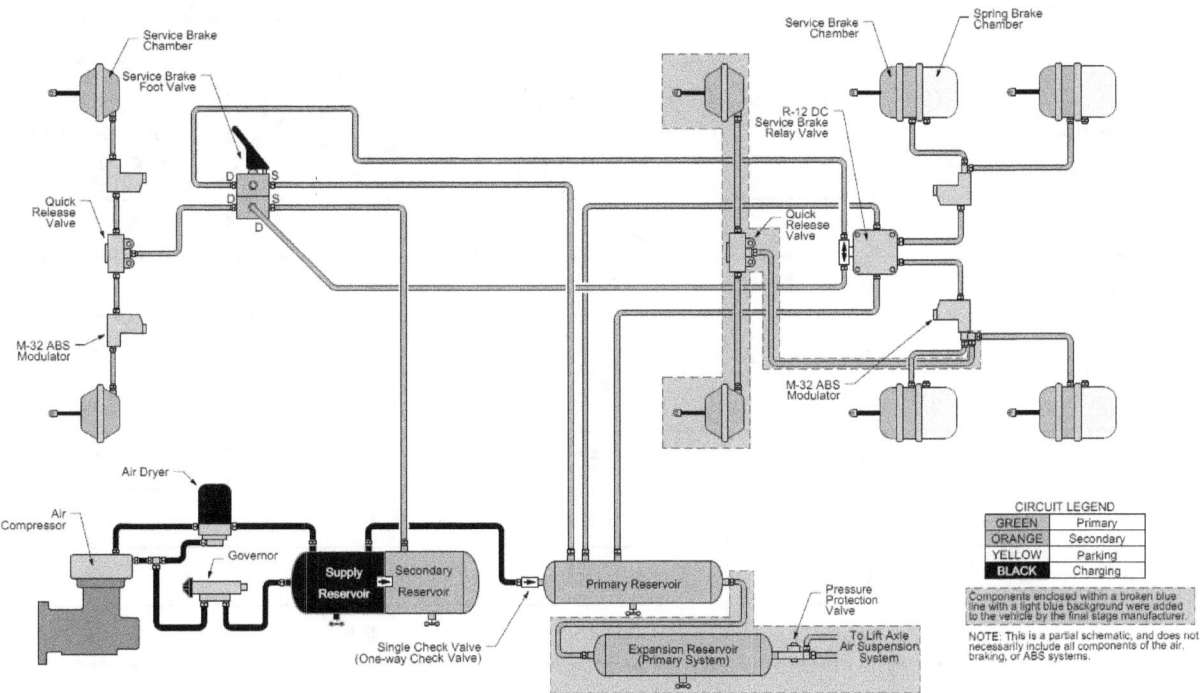

Figure 11. Truck air system schematic.

In 2003, FMVSS 121 required that the air pressure in each brake chamber of new commercial vehicles be capable of reaching (applying) 60 pounds per square inch (psi) within 0.45 second. In addition, FMVSS 121 required that the air pressure within each brake chamber be capable of dropping (releasing) from 95 to 5 psi within 0.55 second.[42] Air brake timing tests were conducted on the truck to measure the time it took for compressed air from the braking system to travel to each axle and reach 60 psi in the brake chambers once the brake pedal was depressed, and once the pedal was released the air would need to drop to 5 psi in each chamber. The axle 2 left and right sides both failed the apply and release tests, axles 3 and 4 left and right sides both failed the release test, and axles 3 and 4 left side failed the apply test. (See table 4.)

---

[42] See 49 CFR 571.121 S5.3.3.1(a) and 49 CFR 571.121 S5.3.4.1(a).

Table 4. Summary of truck air brake test results postcrash (shaded boxes indicate failed times).

Channel	FMVSS 121 Apply Time Required (seconds)	Truck Average Apply Time (seconds)	FMVSS 121 Release Time Required (seconds)	Truck Average Release Time (seconds)
Axle 1 left	0.45	0.324	0.55	0.383
Axle 1 right	0.45	0.322	0.55	0.377
Axle 2 left	0.45	0.613	0.55	0.699
Axle 2 right	0.45	0.627	0.55	0.725
Axle 3 left	0.45	0.583	0.55	0.619
Axle 3 right	0.45	0.440	0.55	0.560
Axle 4 left	0.45	0.580	0.55	0.621
Axle 4 right	0.45	0.433	0.55	0.562

A pressure protection valve is located between the expansion air reservoir and the lift axle suspension and PTO air supply lines. Its purpose is to maintain sufficient air pressure for brake system operation in the case of a failure or pressure loss in one of the accessory systems.[43] NTSB investigators tested the pressure protection valve by introducing a "failure," or leak, in the air suspension system at the supply port to the left air spring/suspension cushion. The secondary system pressure held, providing sufficient pressure for operation of the steer axle brakes, while the primary system pressure—responsible for providing air pressure for the tire and lift axle brakes—drained to 0 psi.

1.6.4 Roll-Off Container Loading and Weight

Herman's Trucking was contracted to provide a roadway construction site with empty roll-off bin containers and to haul away and recycle the contents. The truck driver would pick up a filled container at the construction site and return with a replacement container. The contents of the containers were recycled or disposed of at the Herman's Trucking facility,[44] which was located about 9 miles from the construction site. The loaded vehicle was not weighed at the construction site. To comply with New Jersey law and the New Jersey Department of Transportation (NJDOT) prohibition on overweight vehicles transporting material on the

---

[43] According to 49 CFR 393.207(f), "The air pressure regulator valve shall not allow air into the suspension system until at least 55 psi is in the braking system." This valve, also known as a pressure protection valve, should also cause a pressure of at least 55 psi to be maintained in the braking system in the event of a downstream air leak.

[44] Herman's Trucking is registered with the New Jersey Environmental Protection Agency as a class B recycling facility for construction and landscaping materials.

roadways, the driver was required to go a weigh station located 2.5 miles from the site.[45,46] However, instead of stopping at the weigh station, the driver drove directly to the Herman's Trucking facility and then had the bin container weighed.

Postcrash, the NTSB examined nine company weight slips and found that five showed the truck and load being over the 80,000-pound limit.[47] In March 2012, as a result of the crash, the NJSP issued citations to Herman's Trucking after a determination that the truck was being operated in an overweight condition as measured by several parameters, including statutory limits, axle weight rating, and tire weight rating.[48] (See table 5.)

The Federal-Aid Highway Act of 1956 established the first federal truck size and weight limits. Weight limits on the National System of Interstate and Defense Highways are specified in Federal Highway Administration (FHWA) regulations at 23 CFR 658.17. The Federal-Aid Highway Amendments of 1974 increased the weight limits to: single axle at 20,000 pounds, tandem axle at 34,000 pounds, and maximum gross vehicle weight at 80,000 pounds.[49] On noninterstate highway roads, states can adopt these maximum weight limits or set their own.

Tire manufacturers are required by federal regulation to place the weight rating and inflation pressure on the outside wall of tires. Each vehicle is manufactured with specific tire size, weight capacity, and recommended inflation pressure (shown as psi) to accommodate the vehicle suspension and axle rating specifications. Title 49 CFR 571.119 and 393.75(f) specify the number of truck tires and their weight ratings for the weight anticipated to be transported.

---

[45] According to *New Jersey Statutes*, section 39:A:LL-3, when a vehicle is operated on a noninterstate system highway, the weight limits are as follows: single axle at 22,400 pounds (steer axle and single axles with a distance between consecutive axles greater than 96 inches), tandem axle at 34,000 pounds (distance between axles of 40-96 inches), and total gross vehicle weight limit at 80,000 pounds. When a vehicle is operated on a road designated as part of the interstate highway system, the weight limits are stated in 23 CFR 658.17 and in the weight chart in *New Jersey Statutes*, section 39:3-84(b)(5). Of significance are certain distinctions between federal and state requirements: (1) the federal rule allows for enforcement of the interior consecutive axle weight limits (vehicle bridge rule); and (2) the federal single axle weight limit, at 20,000 pounds, is lower than New Jersey's. State law provides that if the weight being carried exceeds the GAWR certified by the manufacturer for the axle, even if it is less than the maximum weight permitted by New Jersey, the vehicle can be in violation. With regard to the state statutory weight limits, Herman's Trucking failed to meet the exceptions that would have permitted lawful movement of the truck on the highway.

[46] Permanent weigh stations are located at Interstate 78 eastbound and westbound, at mile marker 3.0; Interstate 80 eastbound, 1 mile east of the Pennsylvania state line; Interstate 287 northbound, at mile marker 9.0; and Interstate 295 northbound, at mile marker 3.6, Carney's Point. See www.state nj.us/transportation/-freight/trucking/requirements.shtm, accessed July 1, 2012.

[47] Six slips were dated January 5, 2012; two, January 31, 2012; and one, February 2, 2012. The contract began in November 2011, and one weight slip was found for that month. A tenth slip, for January–February 2012, was illegible.

[48] *New Jersey Statutes*, sections 39:3-20, 39:5B-32, and 39:3-84b(3).

[49] The legislation also established limits to control the amount of weight a vehicle imposes on highway bridges to prevent excessive wear and damage to the interstate highway bridge system road surfaces and bridge structures.

Table 5. Summary of truck axle and tire weight postcrash.

Component	Postcrash Operating Condition (pounds)	Statutory or Manufacturer Specification (pounds)	Overweight Amount (pounds) or Percent of Rated Load	Source[a]
Registered weight	84,950	80,000	4,950 (110%)	Statutory limit
Axle 1	15,100	18,000	84%	GAWR
Axle 1 left tire	7,700	9,370	82%	Tire mfr
Axle 1 right tire	7,400	9,370	80%	Tire mfr
Combination axles 2, 3, 4[b]	69,850	56,400	13,450 (124%)	Statutory limit
Axle 2	12,800	20,000	64%	GAWR
Axle 2 left tire	6,300	9,370	67%	Tire mfr
Axle 2 right tire	6,500	9,370	67%	Tire mfr
Axle 3	29,000	24,680	4,320 (118%)	GAWR
Axle 4	28,050	24,680	3,370 (113%)	GAWR
Axle 3 left tires capacity (dual)	14,300	12,615	1,685 (111%)	Tire mfr
Axle 3 right tires capacity (dual)	14,700	13,220	1,480 (111%)	Tire mfr
Axle 4 left tires capacity (dual)	14,250	12,010	2,240 (119%)	Tire mfr
Axle 4 right tires capacity (dual)	13,800	12,010	1,790 (115%)	Tire mfr

[a] GAWR = gross axle weight rating; mfr = manufacturer.

[b] Axle 2 is the lift axle added by AWE. Axle 3 is the forward drive axle, and axle 4 is the trailing drive axle—both of which were part of the chassis as built by Mack Trucks. Tire location includes dual-mounted configuration of inner and outer tires, except on axle 2.

1.7 School Bus Vehicle Information

1.7.1 Postcrash Inspections

The 2012 54-passenger IC Bus, LLC, model PB01500 school bus was equipped with an International MaxxForce DT 215-horsepower, electronically controlled diesel engine; an Allison model 2500 PTS5-SP automatic transmission; and an electronic control module (ECM).[50] The school bus was also equipped with a Meritor WABCO ABS and hydraulic disc brakes on all four wheels. The odometer reading at the time of the crash was 13,419 miles.

All major mechanical systems were examined, including the steering, suspension, and braking systems. No damage was noted to any of the steering system components; all connections were solid and free of wear or excess play. The brake system was examined and found to be functional and undamaged, passing all system checks; the brake disc rotors and pads were found to be within specified wear limits. The tires were all of the same size—as specified by the manufacturer—and tread depths were found to exceed minimum requirements.[51]

A single-page form generated by GST documented the preventative maintenance performed on the school bus on November 26, 2011; January 18, 2012; and February 3, 2012. This form also documented the quarterly state inspection performed on February 3, 2012. One GST maintenance request form was included in the records, and it indicated that the check engine light and traction control lights were illuminated. Comments on the form indicated that the bus was sent back to the dealership for warranty repairs.[52]

An NTSB recorder specialist examined the ECM and other vehicle modules in the school bus. However, none of the modules on the bus had data recording capabilities, nor was the bus equipped with a video camera or GPS unit.[53]

---

[50] The purpose of the ECM is to control engine timing and fuel injection based on engine and sensor inputs. The ECM is also capable of diagnostics associated with engine or sensor faults, which may then illuminate warnings on the dash. The ABS system on the school bus also had an ECM.

[51] Tread depth is to be measured in two adjacent tread grooves at any location on the tire, not to exceed 4/32 inch for the steer axle and 2/32 inch for all other axles. (See 49 CFR 393.75 [Tires].)

[52] Navistar, Inc., records indicated that eight claims were filed for maintenance covered under the original vehicle warranty. None of the warranty claims included any major engine, driveline, braking, steering, or suspension system components. One warranty claim filed in December 2011 pertained to a bad accelerator pedal sensor. The sensor was causing the vehicle to stall or experience a low power situation when the accelerator was depressed to full throttle (100 percent). The accelerator pedal assembly was replaced. All of the covered warranty work was completed in July–December 2011.

[53] GST had equipped 75 school buses with GPS units. At the time of the crash, its reported goal was to equip all 242 buses in its fleet with GPS units within three years. Sixty-five buses were equipped with visual and audio cameras to record passenger activity; however, these cameras are installed only at the request of each individual school district for specific buses.

1.7.2 School Bus Damage

The primary mechanism of physical damage to the school bus occurred when the truck struck the bus on its driver side behind the rear axle. The damage measured about 74 inches in width and spanned a vertical height from the bottom of the bus body upward about 47 inches. The interior passenger compartment was penetrated at a height of 48 inches from ground level and 65 inches rearward from the rear axle. Postcrash maximum crush intrusion into the passenger compartment measured 17 inches. Red paint transfers from the truck onto the side of the bus extended vertically 67 inches from ground level, and yellow paint transfers from the truck's snowplow mount extended vertically 71 inches.

The school bus was equipped with 11 passenger area windows on the left and right sides.[54] The lower glazing panel in window 8 on the left side was broken just forward of the truck contact damage area. The floor was buckled upward at rows 7–10, with a maximum height of 10 inches. A 12- by 12-inch hole was located in the bus sidewall on the driver side at row 9. (See figure 12.)

Figure 12. Damaged left side of school bus.

---

[54] A small window was located by the driver seat. Counting from front to rear, the left side window 6 and the right side window 5 were emergency exit windows.

After being struck by the truck, the school bus then rotated and struck the traffic beacon support pole. A secondary area of damage was located on the right side of the bus behind the rear axle. This damage spanned a width of 30 inches and extended vertically from the bottom edge of the bus body up to the roofline. The intrusion crushed inward to a maximum penetration of about 10.4 inches (at seating row 8), near the roofline. On the right side, both the upper and lower glazing panels in window 9 and the lower glazing panel in window 8 were broken out in the area of contact. (See figure 13.)

Figure 13. Damaged right side of school bus.

The school bus body was also damaged behind the rear axle, bending the bus body to the right with forward displacement on the chassis frame. The rear emergency door and frame were damaged, but the door was operational postcrash. Interior damage was primarily in the rear of the bus on both the left and right sides. The interior roof was deformed downward over 5 inches above rows 8–10 on both sides. The floor was also damaged in the same region, with maximum vertical deformation of about 10 inches.[55]

---

[55] The interior and exterior of the school bus were scanned with a FARO Focus 3D laser scanner. Details of the scanned data can be found in the Forensic Research and Evidence Documentation Group factual report in the NTSB public docket for this crash.

Each school bus seat shared a common base frame, attached to the floor via a forward and aft leg with anchor at the aisle position and to the sidewall via a forward and aft anchor at roughly the level of the seat pan. The left side bench seats measured 45 inches wide, while the right side bench seats measured 30 inches wide; all passenger seatbacks measured 28.5 inches in height. All seat pans were attached postcrash. However, the interior body sidewall, floor, and some seats were displaced due to the intrusion, including the left side row 9, where the seat pan was displaced laterally approximately 4 inches into the aisle[56] and the seatback was deformed rearward into row 10. The right side row 8 forward wall anchor was sheared off, and the forward and rearward floor posts were bent laterally inward at the sidewall intrusion on the right side of the bus. The right side row 9 seat was shifted laterally into the aisle almost 9 inches. (See figure 14 for interior damage.)

Figure 14. Forward view of interior damage to school bus, rows 8–10.

1.7.3 Acceleration Testing

On February 21, 2012, an exemplar 2012 IC Bus, LLC, school bus was brought to the BCR 528–660 intersection. Tests were conducted to determine the acceleration capabilities of the bus from a stop; the time it takes for the bus to travel from its assumed stopping location to the AOI; and the speed of the bus at the time of collision. Based on the school bus driver's statement, the statement of the exemplar school bus driver, and the results of a traffic observation

---

[56] The seat pan shift was a result of the sidewall deformation. At the time of examination, the seat pan was in direct contact with the bus sidewall, though still attached to the seat frame.

study, the stopping location was estimated to be 16 feet south of the BCR 528 travel lane and 14 feet north of the white stop line on BCR 660. The distance from the school bus stopping location on BCR 660 to the AOI on BCR 528 was 54.4 feet, measured from the bus front bumper. Four timed tests were completed, with the school bus driver directed to accelerate forward at what he considered to be a "normal" rate. The average acceleration rate for the four tests was determined to be 0.14 g. At this average rate, it would take the bus 4.96 seconds to travel from its stopped position to the AOI. The approximate speed based on this acceleration rate is 14.9 mph.

1.7.4 Vehicle Dynamics Study

NTSB investigators conducted a vehicle dynamics study to estimate the speed of the truck at impact with the school bus, and the crash pulse sustained by the bus as a result of the collisions with the truck and the traffic beacon support pole. To achieve these objectives, a series of numerical computer-based simulations were conducted with m-smac software, a commercially available vehicle dynamics software commonly used in the reconstruction of highway truck accidents. It is based on the Simulation Model of Automobile Collisions (m-smac; formerly SMAC) algorithm, which was originally developed by CalSpan Corporation for use in the reconstruction of automobile accidents.

The m-smac (and SMAC) algorithm models forces between vehicles as a series of linear springs and as a function of crush width and depth. The m-smac software uses an "open-forum" form of reconstruction procedure in which the user specifies the dimensional, inertial, crush, and tire properties of the vehicles; initial speeds; impact angles; and driver control inputs. The program then calculates the motion of the vehicles and produces detailed time histories of the vehicle collisions, including collision responses. The user compares the m-smac-projected trajectories and collision deformations to determine the degree of correlation.

Key vehicle parameters used in the simulations—including the simulated vehicle dimensions, center of gravity, and inertias—were estimated based on manufacturer's data, NTSB measurements, the weight of the vehicles, and the weight of the occupants. The data used to model the roadway and the intersection were based on three-dimensional survey data taken by NTSB investigators using the FARO laser scanner. These data included postimpact tire marks, indicating the path of the school bus from the AOI with the truck until the collision with the traffic beacon support pole. Also included in the survey were precollision and postcollision tire marks from the truck, indicating its path from before impact until final rest. Tire mark evidence collected during the scene documentation was sufficient to reconstruct the paths of the vehicles.

Once the vehicle data and scene information were entered into the simulation, an iterative process was used to calculate the motion of the vehicles by varying the impact speeds. The postimpact trajectories and the postimpact speed of the truck were then compared with the available physical evidence. A wide range of possible impact speeds were considered. The simulations that most closely matched the physical evidence indicated that the truck was traveling 40–45 mph at impact and the school bus was traveling 12–19 mph at impact. The calculated speed of the exemplar bus was consistent with the acceleration testing of the bus, which indicated an impact speed of 14.9 mph.

The simulation results also indicated that the crash forces were greatest at the back of the school bus, where the fatally injured occupant was seated. Occupant motion was further studied using MADYMO software, a product of TASS International. (See section 1.13.2.)

1.8 Highway Information

1.8.1 BCR 528–660 Intersection

BCR 660, at the south end of the intersection, is a two-way, single-lane paved asphalt roadway running north–south, with a double yellow pavement stripe designating no passing zones. On the eastbound approach to the intersection, BCR 528 lanes are also separated by a double yellow pavement stripe. BCR 528 lanes past the BCR 660 intersection have a spaced yellow pavement stripe to allow passing in the eastbound direction. The intersection has no dedicated turning lanes for either of the roads. The travel lanes on both BCR 528 and BCR 660 are 11 feet wide and on the approach to the intersection have 5-foot-wide paved shoulders; the shoulders are delineated from the travel lanes by solid white pavement stripes. Both roads had a posted speed limit of 45 mph.

1.8.2 Traffic Control Devices

Both northbound and southbound traffic on BCR 660 was controlled by a STOP sign and a white stop line at the intersection with BCR 528. In July 2008, an overhead flashing intersection traffic control beacon was installed at the intersection; flashing yellow beacons control traffic on the major road (BCR 528), and flashing red beacons control traffic on the minor road (BCR 660).

A flashing yellow beacon requires that vehicular traffic on approach to an intersection cautiously enter the intersection to proceed straight through, to turn right or left, or to make a U-turn, except as such movement is modified by lane-use signs, turn prohibition signs, lane markings, roadway design, separate turn signal indications, or other traffic control devices.

A flashing red beacon requires that vehicular traffic on approach to an intersection stop at a clearly marked stop line; the right to proceed is subject to the rules applicable after making a stop.[57] The *Manual on Uniform Traffic Control Devices* (MUTCD [FHWA 2009])[58] specifies that—in the absence of a marked crosswalk—the stop line should be placed at the desired stopping point. This stopping point should not be located less than 4 feet or more than 30 feet from the nearest edge of the intersecting travel way. The white pavement marking stop line on BCR 660 was located 29.75 feet from the white shoulder line of BCR 528 east. The Burlington County Engineers Office (BCE) indicated that the rationale for placing the stop line further back from the intersection was to provide the needed turning radius for larger vehicles. Figure 15 depicts the turning radius required by larger vehicles turning from BCR 528 onto BCR 660; had

---

[57] According to the *Uniform Vehicle Code*, section 11-403(b), if there is no stop line, the driver must stop at the point nearest the intersecting roadway where he or she has a view of approaching traffic before entering the intersection; if a crosswalk is present, however, the driver must stop before entering the crosswalk.

[58] The MUTCD defines the standards used by road managers nationwide to install and maintain traffic control devices on all public streets, highways, bikeways, and private roads open to public traffic.

the white stop line been placed closer to the intersection, stopped traffic on BCR 660 could have interfered with or been struck by larger turning vehicles.

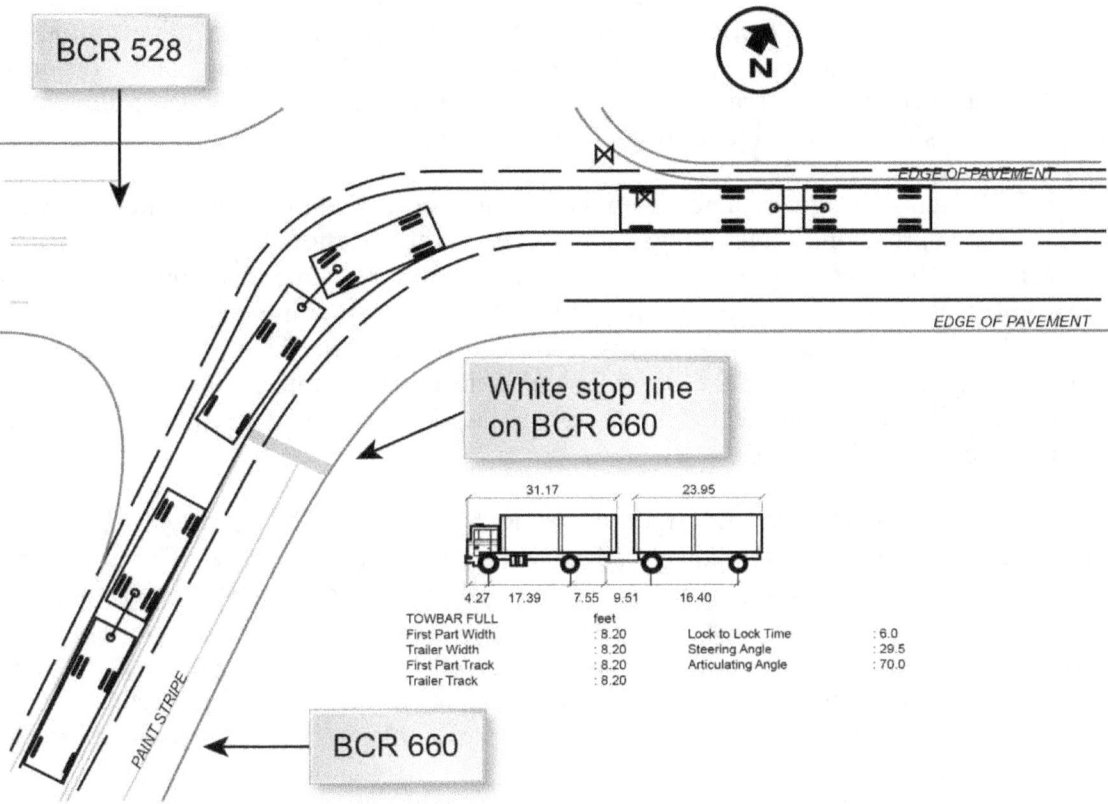

Figure 15. Engineering sketch of BCR 528–660 intersection, showing required turning radius for larger designed vehicles. (Courtesy of Burlington County Engineers Office)

1.8.3 Postcrash Stopping Sight and Departure Sight Distance Testing

Driver line of sight refers to a straight line from the driver's eye toward another object, whereas *vehicle sight distance* refers to the distance from which an approaching driver can see a vehicle (but not necessarily the driver) along the oncoming vehicle's path of travel. The sight distance recommended by the American Association of State Highway and Transportation Officials (AASHTO)[59] for the BCR 528–660 intersection was the stopping sight distance required by the truck driver on the major road (BCR 528) because he was provided preferential right of way by the flashing yellow intersection control beacon.

AASHTO's stopping sight distance is a measurement of the length of the roadway ahead that is visible to the approaching driver along the travel path—that is, how far a driver can see to ensure that there is sufficient distance for a vehicle to stop before reaching a stationary object in its path[60] (AASHTO 2011, 3-2). The AASHTO-recommended stopping sight distance was 360 feet (AASHTO 2011, 3-4, table 3-1)—that is, a truck traveling at the 45-mph design speed at this location, with a reaction time of 2.5 seconds and a deceleration rate of 0.34 g, should be provided a stopping distance of 360 feet if the driver sees a vehicle pull out in front of his or her path from the minor road (BCR 660).

Pine trees were located to the right of BCR 528 eastbound, on private property, approximately 23 feet from the travel lane (1–2 feet outside of the right of way). The trees were located at 10–11 foot intervals, the trunks measured 4–8 inches in diameter, and the tree canopies extended outward about 13 feet and encroached vertically over the right of way. (See figure 16.) Because the trees and their year-round green canopies could limit driver line of sight, vehicle sight distance testing (stopping sight and departure sight testing) was conducted at the intersection at 9:00 a.m. five days after the crash using an exemplar school bus and an exemplar truck.

---

[59] AASHTO represents highway and transportation departments in the 50 states, District of Columbia, and Puerto Rico. It sets standards for all phases of highway system development, to include the design and construction of highways and bridges. The BCR 528–660 intersection was designed and surveyed in 1940, under guidance set by the American Association of State Highway Officials (AASHTO's predecessor). The intent of AASHTO policy is to provide guidance on the design of new projects; the fact that existing streets or highways may not satisfy current AASHTO design values does not mandate the initiation of improvement projects nor does it imply that these roads are unsafe.

[60] The BCR 528 approach to the intersection was essentially level and straight.

Figure 16. School bus driver sight distance (206 feet) from white stop line on northbound BCR 660.

Although stopping sight distance was the required design control at the BCR 528–660 intersection, according to New Jersey requirements and AASHTO policy, departure sight distance is a recommended and desirable feature[61,62] (AASHTO 2011, 3-4, table 1). Departure sight distance is the distance the school bus driver needed to see down BCR 528 to decide when to move through the intersection (AASHTO 2011, 2-3, table 2-1a). (See appendix C.) If the available sight distance for an entering or crossing vehicle is at least equal to the appropriate stopping sight distance for the major road, then both drivers have sufficient sight distance to anticipate and avoid collisions (AASHTO 2011, 9-29). Therefore, both stopping sight and departure sight were included in the vehicle sight distance testing. Sight triangles[63] were measured for the northbound approach of BCR 660 at various increments from the white shoulder line (edge line) of the eastbound BCR 528 travel lane in accordance with AASHTO

---

[61] *Stopping sight distance* is defined as the sum of two distances: (1) the distance traversed by the vehicle from the instant the driver sees an object necessitating a stop to the instant the brakes are applied, and (2) the distance needed to stop the vehicle from when the brake application begins. AASHTO guidelines state that the recommended stopping sight distances are based on passenger car operation and do not explicitly consider truck operation because a truck driver is able to see substantially further beyond vertical sight obstructions than a passenger car driver because of the higher position of the driver's seat. The greater sight distance is considered to balance the greater distance the larger and heavier truck needs to come to a stop from the same given speed as the smaller and lighter passenger car.

[62] Departure sight distance provides sufficient sight distance for a stopped driver on a minor road approach to depart from the intersection and enter or cross a major road. The recommended dimensions of the clear sight triangle for desirable traffic operations where stopped vehicles enter or cross a major road are based on assumptions derived from field observations of driver gap-acceptance behavior (AASHTO 2011).

[63] A sight triangle is a specified area along intersection approach legs and across their included corners that should be clear of obstructions that might block a driver's view of potentially conflicting vehicles. These specified areas are known as clear sight triangles (AASHTO 2011, 9-29).

---

standards for a single-unit truck, which is the design vehicle length most closely related to the school bus[64] (AASHTO 2011, table 2-1a). (See appendix C.)

The exemplar school bus was positioned at seven locations to determine: (1) the stopping sight distance available for a driver in an exemplar 2011 Mack truck to see the stopped school bus while approaching the intersection, and (2) the departure sight distance available for a driver seated in the school bus to observe the approaching truck. The positions of the school bus ranged from distances of 29.75–10 feet from the south edge of BCR 528. Sections 1.8.3.1 and 1.8.3.2 discuss the results of sight distance testing for the school bus positioned at the white stop line and for the bus positioned 16 feet from the south edge of BCR 528, the location that investigators believe was the actual stopping location of the school bus. Appendix C presents the detailed results of the sight distance testing.

1.8.3.1 Truck Sight Distance Results. The exemplar school bus was positioned on BCR 660 at the white stop line, which was 29.75 feet from the edge of the truck's travel lane on BCR 528. At this location, the exemplar truck driver could observe the school bus front bumper and engine hood 287 feet from the intersection. AASHTO required 360 feet of stopping sight distance along the intended travel path of the truck (AASHTO 2011, 3-2)—not the location at which the bus was stopped on BCR 660 because it was not in the truck driver's travel path. Moving the school bus forward 14 feet past the white stop line and 16 feet from the shoulder edge of the truck's lane on BCR 528—the location where the bus driver estimated that he actually stopped—yields a stopping sight distance for the truck driver of 686 feet.

1.8.3.2 School Bus Departure Sight Distance Results. The exemplar school bus was positioned in the same sight triangle locations to determine the departure sight distance for the school bus driver on BCR 660, looking west toward approaching eastbound traffic on the major road (BCR 528). AASHTO (2011) indicates that a time gap of 8.5 seconds should be used for a single-unit truck (the vehicle most similar to the school bus) intending to complete an intersection crossing maneuver.[65] Using that time and the 45-mph speed limit, approximately 565 feet of sight distance was needed. Had the school bus driver stopped at the white stop line, he would have had a departure sight distance of 206 feet. At the location most representative of where the school bus driver stated that he had stopped, he would have had a sight distance down BCR 528 of 463 feet. (See figure 17.)

---

[64] Departure sight distance charts in chapter 9 of the AASHTO policy guide list only vehicles, single-unit trucks, and truck-tractor semitrailers (AASHTO 2011).

[65] The gap time of 8.5 seconds is estimated from when the front of a design straight truck pulls forward from the AASHTO stopping point, which is 10 feet back from the major road edge, until the rear of the design vehicle clears both lanes of traffic.

Figure 17. Sight distance from likely school bus driver stopping position on northbound BCR 660.

The BCR 528–660 intersection is at an acute angle of 63 degrees. A right-skewed intersection can affect a driver's line of sight as well as a vehicle's recommended departure sight distance (AASHTO 2011, table 9-5, figures 9-17 and 9-22) and results in a longer path to travel across the major road (BCR 528). AASHTO design guidelines indicate that though 90 degree intersections are desirable, intersections with a skew angle of 60 degrees offer similar good qualities. An intersection would not have to be considered for realignment construction unless the acute angle was less than 60 degrees (AASHTO 2011, 9-27; Gattis and Low 1997).

1.8.4 Modifications to BCR 528–660 Intersection

In 2006, the posted speed limit on BCR 528 was lowered from 50 to 45 mph in conjunction with an engineering speed survey and a Board of Chosen Freeholders of Burlington County, New Jersey, resolution[66] in anticipation of an expected land use change caused by the nearby development of 1,200 homes and a 60,000-square-foot retail facility.[67] The BCE reported that there were plans for developing intersection improvements once the residential and retail construction was completed. According to the BCE, a modern roundabout[68] (FHWA 2000) was first considered as a design alternative in 2007.

---

[66] See www.co.burlington.nj.us/Pages/ViewDepartment.aspx?did=87, accessed June 18, 2013.

[67] At the time of the crash, 700 homes had been constructed, but the retail facility was not yet built.

[68] A *roundabout* is a circular intersection. Design and traffic control features include yield control of all entering traffic, channelized approaches, and geometric curvature to ensure that travel speeds are typically less than 30 mph. Aprons can be used in small roundabouts to accommodate large vehicles.

Two weeks after the crash, the BCE conducted an average daily traffic (ADT) and speed study that reported the following:

- BCR 528: ADT of 4,558 vehicles and 85th percentile speed of 48 mph

- BCR 660: ADT of 1,521 vehicles and 85th percentile speed of 53 mph.[69]

Traffic flow was compared to current MUTCD traffic flow charts to determine if it satisfied the warrant for a traffic signal.[70] From 2007 through 2011, 16 crashes occurred at the BCR 528–660 intersection: one in 2007, four in 2008, seven in 2009, four in 2010, and none in 2011. None of the crashes were fatal, but the 2007 crash caused serious injury. The others involved complaints of pain or were property-damage-only crashes. Of the 15 crashes, 14 were right-angle collisions: five involving southbound and westbound vehicles, three involving eastbound and northbound vehicles (matching the Chesterfield crash), and six involving right-angle collisions at different quadrants of the intersection. The traffic flow alone—or in combination with the crash history of the intersection—did not meet the warrants for a traffic signal as prescribed in the MUTCD.[71]

The following modifications were made to the intersection postcrash:

- Larger (36-inch-square) STOP signs with retro-reflectorized posts were added in both directions of BCR 660, along with CROSS TRAFFIC DOES NOT STOP plaques on the signs.[72]

- Fourteen trees were moved from the southwest quadrant of the intersection (and replanted) to improve the line of sight from BCR 660 northbound, looking west onto BCR 528.

- Two trees were removed from the northwest quadrant of the intersection to improve the line of sight from BCR 660 southbound, looking west onto BCR 528.

---

[69] The 85th percentile speed is the speed at which 85 percent of vehicles are traveling either at or below, or 15 percent of vehicles are traveling above.

[70] A *warrant* describes a threshold condition based on average or normal conditions that, if found to be satisfied as part of an engineering study, results in analysis of other traffic conditions or factors to determine whether a traffic control device or other improvement is justified. The fact that a warrant for a particular traffic control device is met is not conclusive justification for installation of the device. The need for a traffic control signal is considered if an engineering study finds that all of the following criteria are met: (a) adequate trial of alternatives with satisfactory observance and enforcement has failed to reduce crash frequency; (b) five or more reported crashes of a type susceptible to correction by a traffic control signal have occurred within 12 months, each crash involving a personal injury or property damage apparently exceeding the applicable requirements for a reportable crash; and (c) vehicles per hour condition is met.

[71] The signal warrant evaluation, traffic analysis, and crash history analysis showed that the characteristics of the BCR 528–660 intersection did not satisfy any of the MUTCD warrants for installation of a traffic signal.

[72] According to MUTCD section 2C.59, a CROSS TRAFFIC DOES NOT STOP (W4-P4) plaque may be used in combination with a STOP sign when "engineering judgment indicates that conditions are present that are causing or could cause drivers to misinterpret the intersection as an all stop way" (FHWA 2009).

Postcrash, the BCE finalized the concept plans to construct a roundabout at the BCR 528–660 intersection. The proposed roundabout will feature approach legs designed to include splitter islands to reduce the speed of vehicles entering the intersection, and roadway lighting designed and installed to meet current NJDOT standards. Burlington County has begun the process of acquiring the additional right of way, and it is anticipated that the roundabout will be completed in 2014.

1.9 Motor Carrier Operations

1.9.1 Herman's Trucking Inc.

Herman's Trucking, headquartered in Wrightstown, New Jersey, is an authorized for-hire and private carrier of building materials, general freight, machinery, and construction and landscaping materials.[73] At the time of the crash, the company had a fleet of 23 drivers and 21 vehicles. According to the Federal Motor Carrier Safety Administration (FMCSA), the company was subject to 25 roadside inspections in the 24 months preceding the crash: 14 vehicle inspections with a 14 percent OOS rate (two OOS violations) and 25 driver inspections with a 0 percent OOS rate.[74] The 2009–2010 national average OOS rate was 20.7 percent for vehicles and 5.5 percent for drivers. At the time of the crash, Herman's Trucking had not exceeded the thresholds for safety measurement system values in any of the seven behavioral analysis and safety improvement categories (BASIC).[75]

The FMCSA conducted compliance reviews on Herman's Trucking in March 2006, January 2001, and October 1996—each of which resulted in a satisfactory rating. Postcrash, the FMCSA (through the NJSP) conducted a compliance review and inspected 17 vehicles, which resulted in one OOS violation (a defective taillight). The company received a conditional rating from this compliance review as a result of its crash rate, which was 2.84. According to appendix B, section II.B (Accident Factor), of 49 CFR Part 385, anything above 1.5 is unsatisfactory.[76] Excluding this crash, Herman's Trucking had been involved in three recordable accidents (one fatal, one injury, and one tow-away)[77] in the preceding 24 months. The FMCSA sent the company a warning letter on October 7, 2011, indicating that its crash indicator BASIC was at an "unacceptable level."

---

[73] US Department of Transportation (USDOT) number 354713 and motor carrier number 292150.

[74] February 17, 2010–February 17, 2012 (crash occurred February 16, 2012).

[75] See ai.fmcsa.dot.gov/SMS/Data/carrier.aspx?enc=pUQtpswTy5R3ycKxBBZfUg, accessed March 22, 2012. The FMCSA uses the seven BASICs to measure safety performance and create monthly Compliance, Safety, Accountability (CSA) scores. The categories are unsafe driving, hours of service, driver fitness, controlled substance/alcohol, vehicle maintenance, hazardous materials, and crash indicator. See www.whatiscsa.com/basics/, June 10, 2013.

[76] Crash rates are a ratio of recordable accidents to the number of million miles driven in one year by all company vehicles. Crash rates below 1.5 are satisfactory. Because this accident factor was the only one receiving an unsatisfactory rating, the overall rating was determined to be conditional. See www.fmcsa.dot.gov for all six factors considered during a compliance review process.

[77] These crashes occurred on February 15, 2012, February 23, 2011, and September 3, 2011, respectively.

1.9.2 Garden State Transport Corporation

GST, of Southampton, New Jersey—an authorized interstate exempt for-hire passenger motor carrier—owned the 2012 IC Bus, LLC, school bus.[78] GST began pupil transportation operations in 1997. At the time of the crash, the company had a fleet of 242 school buses with 239 drivers and operated from four New Jersey locations: Southampton, Robbinsville, Freehold, and Washington. According to GST, it operates approximately 500 school routes for 41 school districts in New Jersey.

School transportation is contracted via a bid process; each contract covers a 12-month period and is renewable annually. GST was awarded the NBCRSD contract on May 17, 2011, for the Chesterfield, Mansfield, Springfield, North Burlington County, and North Hanover school districts.[79]

According to the FMCSA, in the 24 months preceding the crash, the company was subject to six roadside inspections: four were vehicle inspections and six were driver inspections.[80] GST's OOS rate for both vehicles and drivers was 0 percent. The company had a total of seven recordable accidents, six of which resulted in injuries.[81] At the time of the crash, the FMCSA safety measurement system database showed that GST had not exceeded the threshold values in any of the seven BASICs.

---

[78] The company provided occasional out-of-state pupil transportation for school activities—about 25 interstate trips per year. According to 49 CFR 387.27(b)(4), an exempt for-hire motor carrier is one engaged in transportation exempt from economic regulation by the FMCSA under 49 *United States Code* (U.S.C.) 13506; however, "exempt motor carriers" are subject to FMCSA safety regulations. Section 387.27(b)(4) provides an exemption from the minimum levels of financial responsibility for "for-hire" carriers engaged in transporting school children in other than "school bus operations." GST did not have interstate operating authority due to exemptions in 49 U.S.C. 13506(a) and 49 CFR 387.27(b)(4), which excluded its need to obtain operating authority. The company is still subject to the FMCSRs.

[79] The Hanover school district included McGuire Air Force Base and Fort Dix Army Base for grades 7–12 only.

[80] This period covered February 17, 2010–February 17, 2012 (crash occurred on February 16, 2012). See the Safety and Fitness Electronic Records database, www.safersys.org, accessed March 22, 2013.

[81] Title 49 CFR 390.5 defines a recordable accident as an occurrence involving a commercial motor vehicle (CMV) operating on a highway in interstate or intrastate commerce that results in a fatality; bodily injury to a person who, as a result of the injury, immediately receives medical treatment away from the scene of the accident; or one or more motor vehicles incurring disabling damage, requiring them to be transported from the scene by a tow truck or other motor vehicle. Also see 49 CFR Part 385, appendix B.

GST underwent a nonrated FMCSA compliance review in 2001; a rated compliance review on February 25, 2011, in which it received a conditional rating; and a third compliance review on October 6, 2011, in which it received a satisfactory rating.[82,83] The FMCSA (through the NJSP) conducted a postcrash compliance review of GST on February 17, 2012.[84] The company received a satisfactory rating, with some violations noted—though none resulted in a less than satisfactory rating in any of the six rating factors. The MCCRU conducted a postcrash compliance review that resulted in a satisfactory rating.

Also on February 17, 2012, the NJMVC inspected the GST fleet. This inspection was prescheduled and unannounced, and was not directly related to the February 16 crash. All vehicles registered for school transportation are inspected twice a year, and unannounced inspections are performed monthly as part of the Governor's School Bus Safety Task Force.[85] The NJMVC inspection revealed 10 incomplete vehicle inspection records; three falsified driver records; 28 driver medical examiner reports that required updating; one driver without a current medical certificate; and 19 buses with violations (of which 18 were reinspected and 16 placed back in service).

---

[82] FMCSA safety ratings are as follows: (1) *satisfactory*: the motor carrier has in place and functioning adequate safety management controls to meet the safety fitness standard prescribed in 49 CFR 385.5; (2) *conditional*: the motor carrier does not have adequate safety management controls in place to ensure compliance with the safety fitness standard, which could result in occurrences listed in 49 CFR 385.5 (a) through (k); (3) *unsatisfactory*: the motor carrier does not have adequate safety management controls in place to ensure compliance with the safety fitness standard, resulting in occurrences listed in 49 CFR 385.5 (a) through (k); and (4) *unrated*: the FMCSA has not assigned a safety rating.

[83] For the February 2011 compliance review, the company was fined $17,210 for failing to meet the required percentage of employees subjected to alcohol and drug testing. For the October 2011 compliance review, the company was fined $7,270 for failing to pay a civil penalty from the February 2011 compliance review, which resulted in a "cease all interstate operations" order from the FMCSA in July 2011. The carrier violated this interstate OOS order when it conducted a trip to Philadelphia, Pennsylvania, on August 23, 2011. The civil penalty was in litigation as of June 10, 2013.

[84] The FMCSA uses the NJSP Motor Coach Compliance Review Unit (MCCRU) as an enforcement program to impose fines or OOS orders in the state of New Jersey. The MCCRU enforces Title 49 of the CFR as well as Title 39 of the New Jersey motor vehicle code. Vehicles that pass a CVSA level 1 or level 5 inspection (of seven levels total) qualify for the required annual inspection under 49 CFR 396.17. A level 1 inspection includes the driver and vehicle (including components in the undercarriage, such as brake adjustment); a level 5 is a terminal inspection of a vehicle only (driver not present). The NJSP Commercial Carrier Safety Inspection Unit implements and enforces federal regulations governing CMV drivers and related safety equipment over state highways, as well as conducting unannounced school bus safety inspections. The MCCRU is responsible for roadside inspection of buses. The MCCRU conducts compliance reviews of motor carriers that have been involved in a serious or fatal CMV crash, which includes an extensive check of records, equipment, and drivers; in addition, the MCCRU conducts compliance reviews of motor carriers that have failed to maintain an acceptable safety rating. See www.cvsa.org for more information.

[85] See www.state.nj.us/mvc/Inspections/SchoolBus.htm, accessed June 10, 2013.

1.10 New Jersey Pupil Transportation Regulations and Organizational Structure

Pupil transportation in New Jersey is regulated at three levels:

- By the New Jersey Department of Education (NJDOE)

- By the county level of the NJDOE

- By local or regional school district.

The NJDOE establishes the requirements for providing pupil transportation by local school districts and private contractors. County offices of the NJDOE monitor pupil transportation and assist the school districts in formulating and implementing pupil transportation programs, including contracting for transportation services from private contractors.

The NBCRSD is a consolidation and cooperative effort of five school districts[86] to provide a centralized school facility for middle and high school education. The regional high school had its own pupil transportation unit, consisting of 37 buses and 33 full-time and 11 part-time bus drivers. However, because this fleet was insufficient to cover all the necessary routes, the NBCRSD contracted with GST for additional school buses. GST received the contract for the accident route on January 25, 2012.[87] The NBCRSD reviewed GST operations several times a week, through reports of traffic citations or crashes and monitoring of student or parent complaints. The NBCRSD indicated that it had received no complaints on the school bus driver prior to the crash.

1.11 Medical Certification for Commercial Vehicle Drivers

According to 49 CFR 390.5, medical certification is required of all interstate commercial vehicle drivers who operate vehicles that weigh 10,001 pounds or more; carry eight or more occupants, including the driver, for compensation; convey 15 or more occupants, including the driver, for no compensation; or transport hazardous materials requiring placards. A CDL applicant must be evaluated and certified by a medical examiner and then be reevaluated biennially, as specified in 49 CFR Part 391. The medical examiner may certify a driver for less than two years if the examiner believes that the driver's physical condition warrants monitoring. A CDL applicant may visit any medical examiner[88] who is licensed by, certified in, or registered

---

[86] The Chesterfield, North Burlington County, Mansfield, Springfield, and North Hanover school districts are responsible for pupil transportation for grades kindergarten–8 and the regional school district for grades 9–12. The NBCRSD transportation director assisted independent school districts with their transportation needs at the local level.

[87] GST was first issued the contract on January 17, 2012, and then reissued the route one week later, on January 25.

[88] In a 1992 rule, the FHWA (which was responsible for administering federal motor carrier safety requirements until January 1, 2000) amended the FMCSRs to expand the definition of "medical examiner" to allow other health care professionals such as physician assistants, advanced practice nurses, and doctors of chiropractic—in addition to doctors of medicine and doctors of osteopathy, authorized previously—to examine CMV drivers. See 57 *Federal Register* (FR) 33276, July 28, 1992.

with a state to perform physical examinations—which includes doctors of medicine, osteopathy, and chiropractic; physician assistants; and advanced practice nurses.

On April 20, 2012, the FMCSA published the final rule that established a National Registry of Certified Medical Examiners. The National Registry was developed to improve highway safety and driver health by requiring that medical examiners be trained and certified to effectively determine whether a CMV driver's medical fitness for duty meets FMCSA standards. The final rule requires that all medical examiners who conduct physical examinations for interstate CMV drivers: (1) complete specific training on FMCSA physical qualification standards, (2) pass a certification test, (3) register on the National Registry system to become a certified medical examiner, and (4) maintain and demonstrate competence through periodic training and testing.[89]

Certified medical examiners agree to keep their National Registry accounts up to date (licenses, training records, etc.) and to transmit at least monthly the results of all CMV driver exams to the FMCSA.[90] According to the final rule, all medical examiner candidates will undergo initial training and certification testing to objectively measure their qualifications and ensure working knowledge of FMCSA regulations and guidelines.[91] Effective May 21, 2012— with compliance required beginning May 21, 2014—the FMCSA will require that motor carriers and drivers use only those medical examiners on the National Registry and accept as valid only certificates issued by such.

In response to the FMCSA NPRM and request for comments on the National Registry of Certified Medical Examiners,[92] commenters asserted that only physicians (doctors of medicine and doctors of osteopathy), advanced practice nurses, and physician assistants—or only health care providers who are permitted by their states to prescribe medications—should be eligible to be certified and to be placed on the National Registry. On February 18, 2009, the NTSB submitted comments reiterating concerns regarding the inability of most doctors of chiropractic, who may serve as examiners in many states, to assess the possible effects of prescription drugs, nonprescription drugs, and drug interactions on commercial vehicle operators. The NTSB also commented that

> . . . it seems extraordinarily unlikely that any brief training program for examiners would be able to replace formal courses in pharmacology and current prescribing experience to the extent necessary for such evaluations. It is unclear why the FMCSA would choose to permit individuals whose backgrounds would not meet minimum requirements to make appropriate certification decisions to apply for

---

[89] Medical examiners must take refresher training every five years and take the certification test every 10 years to maintain their certification. See 77 FR 24104, April 20, 2012.

[90] See nrcme.fmcsa.dot.gov/medical_examiners.aspx, accessed March 25, 2013.

[91] The guidance for the core curriculum specifications is based on current FMCSA regulations on physical qualifications (49 CFR Part 391), as well as the task list developed in the role delineation study (RDS) completed in April 2007, as described in the notice of proposed rulemaking (NPRM). The RDS is a rigorous methodology regularly employed in the certification and medical fields when developing a valid, reliable, and fair certification test. See nrcme.fmcsa.dot.gov/training.aspx, accessed March 25, 2013.

[92] See 73 FR 73129, December 1, 2008.

inclusion on a registry of qualified examiners. The Safety Board strongly suggests that, as an absolute minimum, only individuals who have prescribing authority be permitted to make certification decisions.

1.12 Weather and Visibility

At the time of the crash, the temperature was 31.8 degrees Fahrenheit, the weather was cloudy, conditions were calm with no wind, and the roadway was dry. Astronomical data from the US Naval Observatory indicated sunrise at 6:51 a.m. NTSB investigators observed that at the time (8:15 a.m.) and place of the crash, the sun would have been obscured by the clouds and would not have been in the westbound field of view of drivers.[93] The school bus driver reported in postcrash interviews that on the morning of February 16, 2012, it was cloudy, it was not raining, and the road surface was dry.

1.13 Occupant Protection in School Bus Crashes

1.13.1 Port St. Lucie, Florida

In another recent school bus crash, which occurred on March 26, 2012, a truck-tractor semitrailer struck a large school bus, also resulting in a severe lateral impact collision.[94] The school bus was operated by the St. Lucie County School District and was traveling westbound on Okeechobee Road (State Highway 70) near Port St. Lucie, Florida, when it was hit by the eastbound truck about 3:45 p.m. The school bus was occupied by the driver and 30 student passengers from Frances K. Sweet Elementary School. The 1998 Peterbilt truck-tractor in combination with a flatbed semitrailer was loaded with sod and occupied by the driver only. At this location, Okeechobee Road was configured as a divided four-lane highway with a speed limit of 55 mph.

The school bus had entered a left-turn-only lane and was preparing to turn across the center median and eastbound travel lanes onto Midway Road. The bus turned in front of the eastbound truck, which collided with the right side of the bus in the vicinity of the rear axle. (See figure 18.) Following the impact, the school bus spun clockwise approximately 180 degrees and came to rest facing Okeechobee Road. The truck departed the roadway onto the grass right of way along the southeast side of the intersection and rolled to the left, coming to rest on its left side, with the trailer coming to rest upside down. Figure 19 provides a scene diagram taken from the Florida Highway Patrol Report, Case FHP712-24-005.[95] As a result of the crash, one student passenger on the bus was fatally injured. The school bus driver and 19 other passengers received injuries of varying degrees. The operator of the combination vehicle refused medical treatment.

---

[93] Data from the Chesterfield Downs weather station (KNJCHEST3) in Chesterfield, New Jersey. The US Naval Observatory reported that at 8:00 a.m. the sun was at an altitude of 11.6° above the horizon at 117.3° east of true north.

[94] Highway accident number HWY12FH008. See www.ntsb.gov/investigations/dms.html.

[95] See attachment 1, Florida Highway Patrol Report, in the NTSB public docket for this crash.

Figure 18. Port St. Lucie, Florida, school bus at final rest position. (Courtesy of Florida Highway Patrol)

Francis K. Sweet is a magnet school that serves children from across the county. The transportation center attempts to limit trip times to less than one hour. The driver was doing a "cover run."[96] He had done this cover run 10–12 times previously, but the crash intersection was not on his normal route. The fatally injured child did not typically travel on this bus route.

The school bus was equipped with an onboard video and audio recording system (video event recorder [VER]) manufactured by Seon Design Inc. The system had four active cameras, which recorded at the rate of 15 frames per second. A total of 55 minutes 39 seconds was recorded on each of the four videos. The videos began prior to the loading of the school children and continued through the bus trip to the point of the collision and following. Over 15 minutes of video was recorded postimpact, including vehicle motion, occupant motion, and the initial response of passersby and emergency medical personnel.

---

[96] A cover run is performed when students from one (nonrunning) bus are distributed onto several other buses for the ride to or from school.

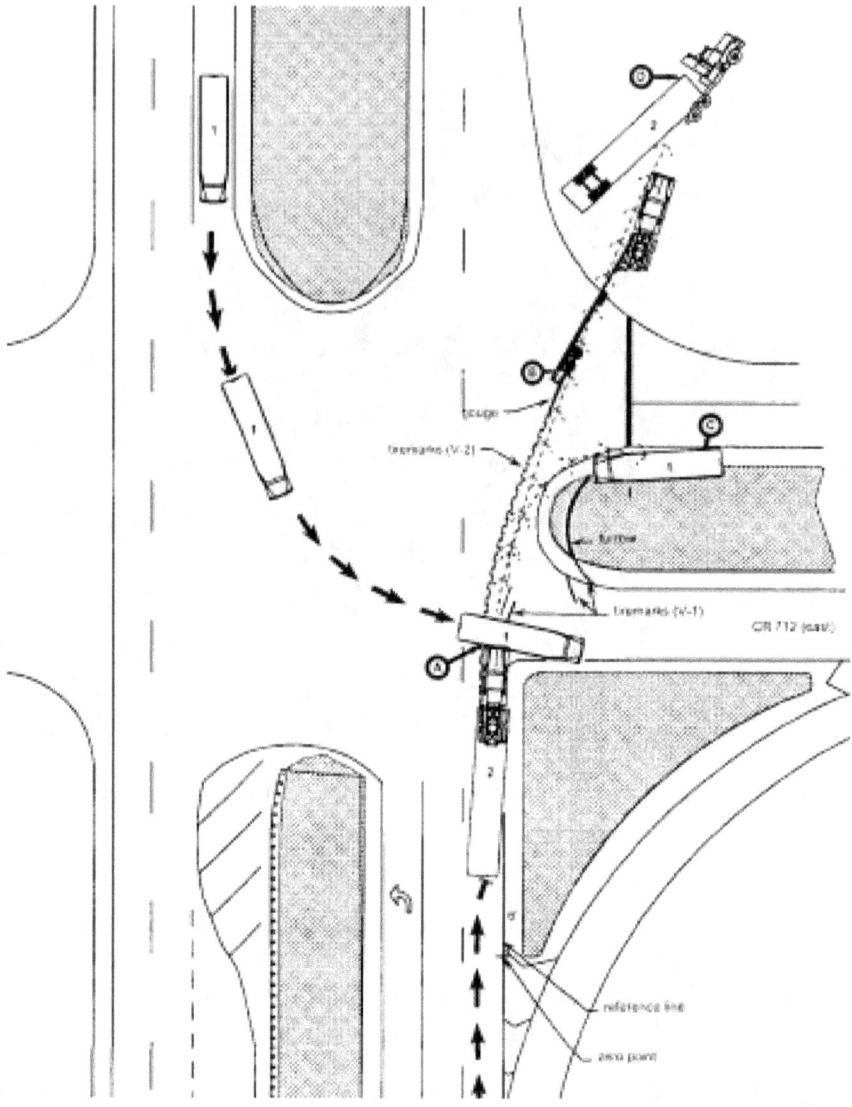

Figure 19. Truck and bus impact sequence, Port St. Lucie, Florida. (Courtesy of Florida Highway Patrol)

1.13.2 Occupant Simulation Studies

Two computer simulation studies (MADYMO, release 7.41) were developed to explore the occupant kinematics during lateral impacts similar to the Chesterfield and Port St. Lucie crashes. The objectives of the Chesterfield study were to create an occupant simulation representative of the crash sequence and to evaluate injury potential to occupants in the rear of the school bus when the truck hit the bus and then when the bus hit the traffic beacon support pole (the first and second lateral impacts). Rear-seated occupants were simulated in various restrained conditions—including unbelted, lap-only belted, and lap/shoulder belted. Similarly, the Port St. Lucie study focused on a representative crash sequence, but additional information from the onboard VER was used to corroborate the simulation. Again, rear-seated occupants were simulated, but the focus was on lap-only belts and lap/shoulder belts to explore the potential benefit of upper body restraint.

2 Analysis

2.1 Introduction

About 8:15 a.m. on February 16, 2012, near Chesterfield, New Jersey, a school bus was stopped approximately 14 feet north of a white stop line on northbound BCR 660 at the intersection with BCR 528. The bus driver stated that he looked left, looked right, and then looked left to check traffic on BCR 528 before accelerating forward to cross the intersection. After traveling northbound for approximately 54.4 feet, the school bus was struck by a Mack roll-off truck traveling east on BCR 528.

Postcrash acceleration testing of an exemplar bus determined that the school bus was traveling approximately 14.9 mph when it was struck. A vehicle dynamics study determined that the speed of the truck at impact was 40–45 mph. New Jersey state law requires that drivers stopped at an intersection proceed forward only after yielding the right of way to all traffic on the intersecting street that is so close as to constitute an imminent hazard. NTSB investigators completed an accident reconstruction (section 2.2) to establish the speed and location of the truck when the school bus driver proceeded forward to cross the intersection, to ascertain whether the truck was within the bus driver's available sight distance of 463 feet, and to determine whether the truck driver could have mitigated the severity of the crash if he had reduced his speed.

The school bus driver's first awareness of the approaching truck as he crossed the intersection was when he felt an impact on the rear driver side of the bus and the "bus went into the air." NTSB investigators analyzed human factors related to the bus driver (section 2.3) to identify possible reasons why he did not see the truck. The analysis considered the bus driver's scanning behavior, fatigue, use of prescription medications, emotional state, a number of medical issues (chronic back and leg pain, depression, and anxiety), and frequent use of alcohol. The bus driver possessed a current medical certificate issued on January 10, 2012. Section 2.3 also evaluates the CDL medical examination process, including whether the bus driver should have been medically certified to operate a CMV; the self-reporting of medical conditions and prescription medications; the lack of any medical followup by the medical examiner; and the qualifications of chiropractors and other professionals without prescription medication prescribing authority to be permitted to make medical certification decisions.

The truck was inspected postcrash and determined to have a number of brake defects difficult to detect, to be overweight, and to have an improperly installed lift axle brake system. Section 2.4 evaluates the effects of these conditions on braking efficiency and the severity of the crash, in addition to discussing how vehicle onboard brake stroke monitoring systems and onboard weighing systems can provide information to the driver regarding defective brakes or overweight conditions.

The BCR 528–660 intersection was skewed at 63 degrees, and the white stop line for traffic on northbound BCR 660 was placed nearly 30 feet from the travel lanes of BCR 528. Additionally, pine trees on the residential property on the southwest quadrant of the intersection were located south of the eastbound BCR 528 travel lane. Section 2.5 evaluates highway issues, including design and sight distance obstructions and their potential effects on the cause of this crash. Section 2.5 also discusses intersection safety and how intelligent transportation systems technology, such as vehicle-to-vehicle (V2V) and vehicle-to-infrastructure (V2I) communications, can be used to prevent future intersection crashes.

The impact of the truck with the school bus and the subsequent impact of the bus into the traffic beacon support pole contributed to one fatality and severe injuries among the school bus occupants. The bus was equipped with lap belts, though some students were either not wearing them or wearing them improperly. Section 2.6 discusses school bus occupant protection issues by evaluating the impact forces in the Chesterfield and Port St. Lucie, Florida, crashes; the benefit of occupant restraints on school buses; the need for improved protection to school bus sidewalls and seat frames; and the need for improved training of bus drivers, students, and parents on the importance of ensuring use and proper fit of school bus seat belts.

The remainder of this section addresses those factors that the NTSB investigated and determined did not contribute to the crash, in addition to discussing the timeliness and effectiveness of the emergency response.

The drivers of both the school bus and the truck were licensed, and both commercial drivers possessed current medical certificates. Each driver was familiar with the BCR 528–660 intersection, and neither was found to be distracted by in-vehicle or external environmental diversions. Neither driver was using a cell phone at or near the time of the crash. Postcrash toxicological test results for both drivers were negative for alcohol and illicit drugs.

The truck driver reported (and his medical certificate and medical records showed) that he was not taking any prescription medicine; and, in his interview he stated that he received about eight hours of sleep in the nights prior to the crash. He did not report any adverse life events that could have affected his driving.

Examination of the school bus revealed no preexisting mechanical defects or deficiencies that would have caused, or contributed to the severity of, the crash. GST, the school bus motor carrier, received a satisfactory compliance review postcrash—the same rating it had received in three previous reviews.

At the time of the crash, the weather was cloudy and the roadway was dry. The sun was not in the forward or westbound field of view of northbound drivers on BCR 660, and glare was not a factor. The NTSB concludes that none of the following were factors in the crash: (1) alcohol impairment or illicit drug use by the school bus driver, or alcohol, over-the-counter, prescription medication, or illicit drug use by the truck driver; (2) in-vehicle or external distractions, including cell phone use; (3) truck driver fatigue; (4) operations by GST, the school bus motor carrier; (5) school bus mechanical defects or deficiencies; or (6) weather.

The initial 911 calls provided an accurate location for the crash, emergency dispatch did not encounter any problems while handling emergency calls, and emergency responders were on scene within minutes. The first emergency call was received at 8:17 a.m.; the dispatcher called for multiple firefighting and EMS units beginning at 8:20 a.m.; and by 8:22 a.m., the fire department had people on scene and setting up incident command. Law enforcement responders were dispatched immediately and began arriving on scene by 8:21 a.m. to provide roadway management. EMS units began arriving on scene within 6 minutes of the first 911 calls, and both on-scene medical care and preparation for transport to hospitals were prompt and efficient. The NTSB concludes that the emergency response was timely and adequate.

2.2 Accident Reconstruction

Based on a vehicle dynamics study, NTSB investigators determined that the truck collided with the school bus at a speed of 40–45 mph. Approximately 100 feet before the AOI, the left side lift axle tire on the truck created a tire friction mark. Applying the results of brake force analyses and air pressure buildup timing, investigators determined that the truck was traveling at a speed of 53–58 mph before the truck driver applied the brakes. Further analysis found that he applied the brakes 176–183 feet—or 2.3–2.5 seconds—before impact. Because acceleration testing determined that the school bus was in motion for approximately 5 seconds before impact, the truck was calculated to be 368–406 feet west of the AOI when the school bus began to move forward.[97]

Despite the presence of trees located west of the intersection and parallel to the BCR 528 eastbound travel lane—which would have intruded into the school bus driver's field of view at the white stop line—he was adamant in his interviews with both law enforcement and the NTSB that he had sufficiently pulled forward of the stop line before entering the intersection, such that he was able to see clearly in both directions. The NTSB conducted departure sight distance tests at the intersection, which indicated that the truck could be seen from as far as 463 feet from where the bus driver had likely stopped prior to entering the intersection. Further analysis established that the braking distance available to the truck driver, once the bus driver began to accelerate into the intersection, was insufficient to effectively avoid the collision. Accordingly, the NTSB concludes that the truck was within the school bus driver's available line of sight and within a hazardous proximity when the bus driver began to cross the intersection.

The truck driver said that he had been traveling about 45 mph, the posted speed limit, as he approached the intersection. This stated speed conflicts with the NTSB-calculated speed of 53–58 mph. Analysis determined that when an initial speed of 45 mph is applied to the crash scenario, the resulting impact is inconsistent with the dynamics of the collision (postimpact vehicle motion and travel distance) and simulation analyses. Had the initial speed of the truck been slower, its speed at impact with the school bus—and consequently the impact energy— would have been substantially reduced, though the crash would likely still have occurred.[98] Therefore, the NTSB concludes that the driver of the truck was driving in excess of the posted speed limit before braking for the impending collision, and this higher speed contributed to the severity of the crash.

---

[97] This range of distances is based on the time it took for the school bus to proceed from its stopping location to the AOI. The time is based on an approximate starting location and the acceleration rate of the school bus.

[98] A reduction in the truck's initial speed to 45 mph was estimated to reduce the impact speed by as much as 35 percent.

As part of accident reconstruction, the NTSB evaluated how problems identified with the truck's braking system, its overweight condition, and the improperly installed lift axle brake system affected braking efficiency and the severity of the crash. This analysis is included in section 2.4. Section 2.3 reviews possible reasons why the school bus driver did not observe the approaching truck before proceeding to cross the intersection.

2.3 School Bus Driver "Fitness for Duty"

The NTSB examined factors that might have resulted in the school bus driver not detecting the oncoming truck and pulling out into the intersection, including the driver's scanning behavior, fatigue, use of prescription medications, emotional state, and medical conditions. The following analysis is based on the bus driver's interview statements, 72-hour activity history, a review of medical and prescription records, and toxicology results.

2.3.1 Scanning Behavior

According to the school bus driver, as he approached the BCR 528–660 intersection, nothing seemed out of the ordinary. He stopped on BCR 660 approximately 14 feet north of the white stop line and 16 feet south of the BCR 528 travel lane. He stated that he could see well to the left and right, and on-scene testing indicated that he had 463 feet of unobstructed sight distance to the west in the direction of the approaching truck. The bus driver demonstrated to NTSB investigators how he looked left, then right, and then left again. He turned his head to the left approximately 90 degrees to show how he looked for traffic from that direction.

The school bus driver could not recall if there was any crossing traffic as he approached the intersection, or if there was traffic on the opposite side of the intersection. He did not provide any details regarding the length of his glances to the left and right. He did state that he "looked left, didn't see anything. I looked right, did not see anything. I looked back to the left, did not see anything. I put my foot on the gas, proceeded to try to proceed through the intersection." The driver did not report whether he continued to scan for oncoming traffic as he entered the intersection.

There were no independent witnesses to verify the extent of the school bus driver's scanning behavior at the intersection. Based on accident reconstruction, it was determined that the truck was 368–406 feet west of the AOI when the school bus began to move forward. It is estimated that the approaching truck was approximately 110 degrees to the left from the heading of the bus at this location, as shown in figure 20. However, it took the school bus over 2.5 seconds to travel the first 16 feet before it reached the edge of the BCR 528 travel lane. During this time, there was no evidence to suggest that the bus driver effectively scanned for oncoming traffic as he crossed the intersection.

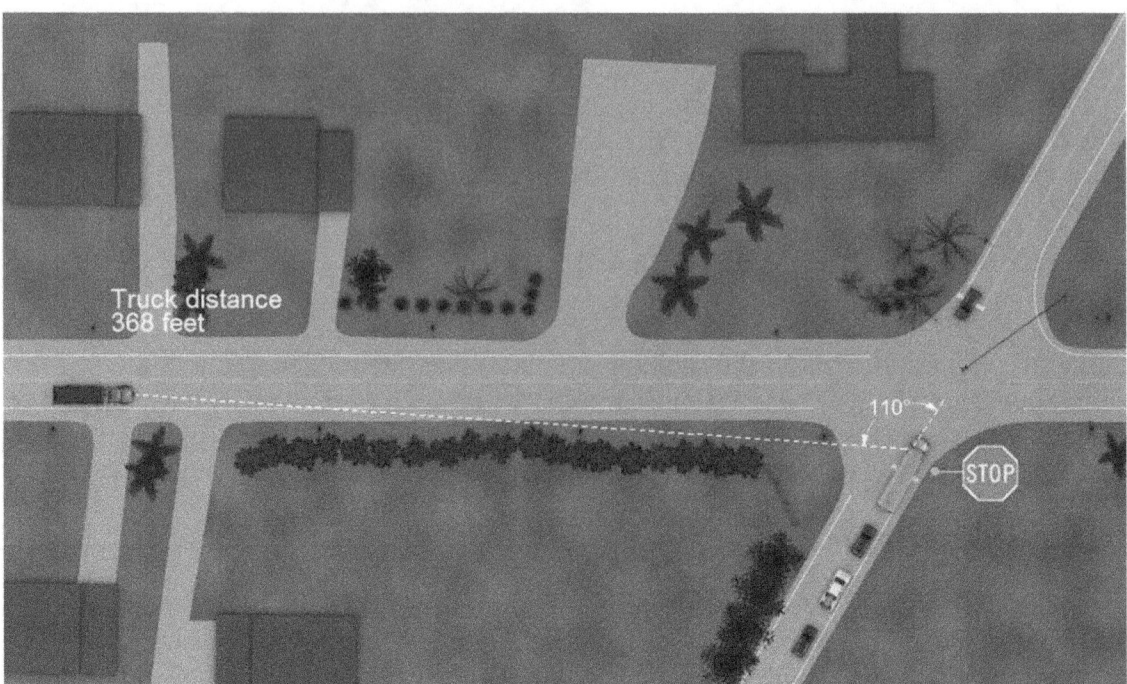

Figure 20. Accident reconstruction image showing approaching truck and school bus at intersection.

Over the entire 5 seconds the school bus was in motion leading up to impact—from when the bus driver completed his left-right-left scanning of the intersection and then looked forward only, he did not observe the approaching truck. The bus driver's first indication of oncoming traffic was feeling the actual impact. In a New Zealand study of intersection crashes, failure to observe oncoming traffic was identified as the most significant causal factor (Land Transport NZ 2005). Safe crossing of an intersection requires visually detecting and monitoring traffic and potential conflicts, along with persistent and accurate scanning of the environment. The NTSB concludes that the school bus driver did not effectively scan BCR 528 for oncoming traffic and failed to observe the approaching truck prior to impact.

The NTSB also considered the possibility that the truck was actually in the school bus driver's field of view when he looked to the left, but he failed to see it. Rumar (1990) identified two important causes of "look but fail to see" errors: sensory limitations and cognitive factors. Perceptually, the truck would have been within the bus driver's view. Cognitively, the bus driver may have been impaired. Potential factors that may have affected the bus driver's cognitive processing—such as fatigue, emotional state, and use of prescription medications—are discussed in detail below.

2.3.2 Driver Fatigue

The NTSB examined several factors to determine if the school bus driver was fatigued at the time of the crash. These factors included quantity of sleep, quality of sleep and any potential sleep disorders, time of day, time awake and on task, and use of prescription medications.

2.3.2.1 Quantity of Sleep. The school bus driver kept a fairly regular sleep schedule, retiring at 11:15–11:30 p.m. and arising at approximately 4:30 a.m. on each of the three days prior to the crash. Although this consistent schedule during the week provided him with at least 5 hours in bed each of the three nights before the crash (excluding times he awoke to use the bathroom), it did not allow for the 7–9 hours needed for a full, restorative rest cycle.[99] When interviewed, the bus driver told investigators that he took a 10–15 minute nap every day at lunch, which would leave him feeling refreshed. He also said that on the weekends, when he did not have to work, he slept approximately 2 hours longer each day.

Although, in general terms, the school bus driver's sleep pattern appears to be stable, the length of his sleep each night is of concern. Performance decrements are seen the day following even low levels of sleep loss—as little as 1 or 2 hours (Wilkinson, Edwards, and Haines 1966; Belenky and others 2003)—and the odds of being in a crash are 2.6 times higher for a driver who had only 6.0–6.9 hours of sleep the night before (Stutts and others 2003). Performance decrements associated with fatigue include slowed response times, loss of situational awareness, and errors in short-term memory (Goel and others 2009). The driver's longer sleep periods on the days he did not have to work are termed "recovery sleep" and indicate that he was not getting the sleep his body required during the week.

The length of sleep an individual needs varies with genetics, circadian timing, and sleep debt, among other factors (Carskadon and Dement 2005). The school bus driver had the opportunity to obtain sleep during weekday nights, but—based on his self-reported time in bed—he received considerably less than even the minimum recommended 7 hours for at least the three nights prior to the crash. Studies have shown that repeated days of only 3–6 hours in bed increase lapses of attention on a psychomotor vigilance task (Van Dongen and others 2003; Drake and others 2001; Dinges and others 1997). A repeated reduction in sleep time results in chronic partial sleep restriction—which is more common than total sleep deprivation and can be caused by medical conditions, sleep disorders, work demands, and social and domestic responsibilities. This chronic sleep debt was evident by the driver's weekend sleep schedule, in which he would sleep for approximately 7 hours each night and wake up naturally, indicating that he was engaging in recuperative sleep, trying to erase the sleep debt he accumulated during the week. Mood changes—including increased sleepiness, fatigue, irritability, difficulty concentrating, and disorientation—are commonly reported during periods of sleep loss (Bonnet 2005).

2.3.2.2 Quality of Sleep. The school bus driver had no history of OSA or other sleep disorders, though he said he had been told that he snores on occasion. Based on his most recent commercial driver fitness exam, the bus driver was 73 inches tall and weighed 216 pounds, which corresponds to a body mass index (BMI) of 28.5. Under FMCSA Medical Review Board

---

[99] The National Sleep Foundation maintains that adults (over age 17) typically need 7–9 hours of sleep a night. See www.sleepfoundation.org/article/how-sleep-works/how-much-sleep-do-we-really-need, accessed May 8, 2013.

recommendations—which call for OSA screening for all drivers with a BMI over 30—the bus driver, despite having a BMI that would classify him as overweight,[100] would not have been indicated as needing to be screened for OSA.

In further reviewing the school bus driver's quality of sleep, NTSB investigators also looked at other medical conditions that could have prevented a restful night sleep. On July 6, November 11, December 16, and December 30, 2011, the bus driver visited his orthopedic doctor and complained of increasing pain in his right hip and lateral aspect of his right leg, along with pain in his lower lumbar (lower back) region that radiated down his thigh. In addition, during the December 30 visit, the doctor noted that the patient was ambulating with a significantly antalgic gait and that his pain was along the SI joints as well as his lumbar spine. The driver had initially been prescribed tramadol for pain and then oxycodone. Toxicology results revealed that tramadol was in his system at the time of the crash, which indicates that he was still suffering from pain that could affect his quality of sleep.

The school bus driver also reported drinking alcohol each evening and waking up at least once or twice per night to use the bathroom within a few hours of going to bed. The onset of sleep may occur quickly after an individual has consumed alcohol; however, alcohol is also disruptive to sleep, causing individuals to wake up frequently during the second half of the sleep period. Alcohol consumption reduces sleep quality and may lead to daytime fatigue and sleepiness[101] (NIAAA 1998).

2.3.2.3 Time of Day. Some systems in the human body fluctuate in cycles of approximately 24 hours; this circadian rhythm can affect body systems, including body temperature, heart rate, blood pressure, and sleepiness, among others (Kroemer, Kroemer, and Kroemer-Elbert 1990). At low points in the cycle, subjective sleepiness is most pronounced and human performance is most degraded. The time of this crash—8:15 a.m.—was outside of the circadian low points.

2.3.2.4 Time Awake and On Task. The length of time awake and on task is also associated with increased risk. The school bus driver had reported that he awoke to his alarm at 4:30 a.m. The crash occurred at 8:15 a.m.—allowing for a length of time awake of less than 4 hours—and the driver's time on task was less than 2 hours. This amount of time awake and on task is unlikely to have contributed to his fatigue.

2.3.2.5 Use of Prescription Medications. The school bus driver's postcrash toxicology specimens were positive for tramadol, 7-amino-clonazepam (the predominate metabolite of clonazepam), and desmethylvenlafaxine (a metabolite of desvenlafaxine).[102] All three of these prescription medications can have sedative effects and contain warnings of other possible side effects.

---

[100] The US Centers for Disease Control and Prevention define a BMI of 25–29.9 as overweight. See www.nhlbi.nih.gov/guidelines/obesity/BMI/bmicalc.htm, accessed May 30, 2013.

[101] See also pubs.niaaa.nih.gov/publications/aa41.htm, accessed May 8, 2013.

[102] The school bus driver stated that he took both clonazepam and desvenlafaxine on the morning of the crash. He awoke about 4:30 a.m. that day and departed for GST at 5:30 a.m. The crash occurred at 8:15 a.m. Therefore, he had taken the medications 2.75–3.75 hours prior to the crash. His blood was drawn for toxicology testing 3.75 hours after the crash, yielding a total possible time difference of 7.5 hours for toxicological results after ingesting the medicines.

Clonazepam—the longest acting of the benzodiazepines (Kreuger and Leaman 2011)—is considered to be a hypnotic drug. It is used for the treatment of anxiety and seizure disorders. Because benzodiazepines have the potential to impair judgment, thinking, or motor skills, patients are cautioned about operating hazardous machinery, including automobiles, until they are reasonably certain that clonazepam therapy does not adversely affect them. An Australian truck driver study showed that drivers who consumed benzodiazepine-type medications were 1.91 times more likely to have had a crash in the previous three years and 2.4 times more likely to have crashed with the use of narcotic analgesics (Howard and others 2004; Kreuger and Leaman 2011). Desvenlafaxine is an antidepressant SNRI that is used to treat depression. This medicine also carries a warning that it may cause drowsiness or dizziness and that those who take it should avoid driving, using machines, or doing anything else that requires alertness.[103]

Tramadol is a centrally acting sedating analgesic used to treat pain. Warnings include impairment of mental or physical abilities required for the performance of potentially hazardous tasks, such as driving a car or operating machinery. At least one study has noted a decrease in the ability to perform complex tasks with use of the drug tramadol (Hummel and others 1996).

During the postcrash interview with the school bus driver, he stated that though some of his medications warned against driving, they did not cause any side effects and he had been on the medications for enough time that he was accustomed to them. Although it is not possible to determine if the bus driver was experiencing impairment from the combination of medications at the time of the crash, possible sedation from the use of clonazepam and tramadol cannot be ruled out.

2.3.2.6 Driver Fatigue Summary. When humans are impaired by fatigue, they are more susceptible to lapses and performance errors related to slowed reaction time (Kleitman 1963; Babkoff, Caspy, and Mikulincer 1991), reduced vigilance (Glenville and others 1978), sustained attention (Dinges 1995), lane tracking ability (Lamond and Dawson 1999), and impaired cognitive processing (Dinges and Kribbs 1991).

The NTSB determined that at least two fatigue indicators most likely affected the school bus driver: quantity of sleep and quality of sleep. In numerous crash investigations, the NTSB has identified the tendency of fatigue to interfere with the ability of operators to redirect their attention. In this crash, the bus driver's reduced vigilance, partially caused by fatigue, negatively influenced his driving performance and was likely exacerbated by his use of sedating medications. These factors reduced his ability to maintain the alertness required for the dynamic driving environment of the BCR 528–660 intersection.

When considering fatigue for the school bus driver, NTSB investigators determined that an operational error occurred involving the driver failing to observe the approaching truck and proceeding directly into its path. As described above, fatigue can result in reduced vigilance and cognitive processing, and such deficits are consistent with the bus driver's actions. Accordingly, the NTSB concludes that the school bus driver was fatigued due to acute sleep loss, chronic sleep debt, and poor sleep quality associated with his medical conditions and alcohol use; the sedative side effects from prescription medications; and the synergistic effect of these factors. The NTSB

---

[103] See www.ncbi.nlm.nih.gov/pubmedhealth/PMHT0009870/?report=details, accessed May 28, 2013.

further concludes that the school bus driver's fatigue contributed to his reduced vigilance and detection of the approaching truck.

2.3.3 Emotional State

Approximately 36 hours prior to the crash, a family member close to the school bus driver passed away. During the postcrash interview with the bus driver, he was visibly upset over the loss of his relative. He was also concerned about finances; he had taken the job as a school bus driver to make extra money. Researchers have found that factors of this kind—grief, loss, and anxiety—can adversely affect driving performance. Practitioners in the medical and driving field recognize that "a person severely ill with anxiety or depression, whose reactions are retarded, who cannot concentrate or make decisions and who is absorbed in worries and problems will not be a safe driver" (Harris 2000). Grieving drivers or those reporting intense mind wandering have experienced cognitive issues while driving (Rosenblatt 2004; Galera and others 2012). An examination of crashes in New Zealand from 2002 to 2003 found that six fatal crashes, 20 serious injuries, and 85 minor injuries could be attributed to drivers who were "emotionally upset–preoccupied" (Gordon 2007).

Despite the possible negative side effects of depression and anxiety on the ability to concentrate, the school bus driver did not self-report that grief and anxiety were on his mind at the time of the crash.

2.3.4 CDL Medical Examination Process

The significant involvement of medications in crashes is demonstrated in the FMCSA Large Truck Crash Causation Study, which showed that prescription medications were considered "accident associated factors" in 26 percent of 967 accidents[104] (FMCSA 2007). The FMCSA, however, does not provide any regulatory or nonregulatory guidance regarding the use of specific medications.[105] The school bus driver was prescribed numerous medications, including sedating medications for chronic low back pain, for mood disorders (anxiety and depression), and to treat alcohol abuse (withdrawal symptoms). The NTSB reviewed the school bus driver's self-reporting of medical conditions, potentially disqualifying medical conditions, and the medical certification of CMV drivers.

2.3.4.1 Self-Reporting of Medical Conditions and Prescription Medications. The school bus driver had received his medical certificate 36 days before the crash. As discussed below, he had reported most, but not all, of his prescription medications and medical conditions to the CDL examining doctor. The driver's personal records show two medical diagnoses (chronic low back pain and alcoholism) that he did not report, along with two prescription medications (tramadol and oxycodone). However, postcrash, it was shown that the driver was aware of his conditions and medications because his hospital records indicate that he self-reported to the emergency

---

[104] The study authors define an accident associated factor as a "factor that could be important"—not the critical reason or the "event" causing the crash.

[105] Disqualifying offenses for commercial drivers include the following: "Driving a commercial motor vehicle under the influence of a 21 CFR 1308.11 Schedule I identified controlled substance, an amphetamine, a narcotic drug, a formulation of an amphetamine, or a derivative of a narcotic drug." See 49 CFR 391.15(c)(2)(ii).

room physician a history of cervical spine chronic pain, anxiety disorder, and depression (but not his history of alcoholism or alcohol abuse),[106] along with all his current medications for these conditions: clonazepam, oxycodone, desvenlafaxine, and citalopram. (He did not report his use of tramadol.)

Commercial drivers in interstate commerce may operate commercial vehicles if: (1) the driver has no established medical history or clinical diagnosis of orthopedic, muscular, or neuromuscular disease that interferes with his/her ability to safely control and operate a commercial motor vehicle; (2) the driver has no mental disease or psychiatric disorder that would interfere with the ability to safely drive a commercial vehicle; and (3) the driver has no current clinical diagnosis of alcoholism.[107]

On January 10, 2012, the school bus driver visited the doctor of chiropractic for his medical certification examination. The examining doctor was not the driver's regular physician, and he had access only to the medical history the driver provided. On the health history section of the examination form, the driver checked "no" for the question on whether he had chronic low back pain, and he did not disclose on the form his use of tramadol or his prescription for oxycodone. He did check "yes" for regular, frequent use of alcohol, and he wrote that he drank two glasses of wine per night. He checked "yes" for a nervous or psychiatric disorder, crossed out "severe" and wrote "mild" depression, wrote that he also had "anxiety," and then listed the medications clonazepam and citalopram. There is no written indication that he disclosed the underlying reason for his family doctor to prescribe these medications—which was a diagnosis first of alcoholism then of alcohol abuse in October, November, and December 2011. The CDL examining doctor noted on the bus driver's CDL form that he discussed the driver's known medications and stated that they did "not interfere with driving."

Medical examiners largely base the certification determination on objective data (physical examination) and driver identification of health history, prescribed medications, and alcohol or illegal drug use on the examination form. The purpose of the health history and physical examination is to detect the presence of physical, mental, or organic conditions that may affect the ability of the driver to safely operate a CMV (Blumenthal and others 2002); however, this determination is dependent on the driver providing an accurate and full health history and the thoroughness of the physical examination. The school bus driver knew that he had chronic low back pain and that he was being treated for alcoholism and alcohol abuse; yet, this disclosure does not appear in the written information he provided or in the examining doctor's notes.

Medical certification depends on a comprehensive medical assessment, including an informed medical judgment about single as well as multiple conditions and their impact on a driver's overall health; however, because medical determinations rely on patient self-reporting, it is not always possible to systematically determine disqualifying medical conditions. The NTSB concludes that the school bus driver failed to disclose pertinent information about his medical

---

[106] The school bus driver also reported high blood pressure, reflux disorder, and hypercholesterolemia.

[107] See 49 CFR 391.4(b)(7). The FMCSA also excludes an established medical history of rheumatic, arthritic, and vascular disease. See www.fmcsa.dot.gov/rules-regulations/administration/fmcsr/fmcsrruletext.aspx?reg=391.41 for additional requirements for physical qualifications to drive a CMV.

history as required on the CDL medical certification examination form, which prevented the accurate assessment of his qualifications to drive a school bus in commercial operations.

2.3.4.2 Disqualifying Medical Conditions. Although the school bus driver failed to disclose his entire medical history on the CDL medical certification examination form, a number of conditions he reported were potentially disqualifying. The medical examiner's review of the bus driver's orthopedic condition, nervous or psychiatric disorders, and frequent use of alcohol are discussed below.

Orthopedic Disease. Eleven days prior to his CDL medical examination, the school bus driver went to his orthopedic doctor, who noted that he was ambulating with a significantly antalgic gait; he underwent an MRI; he was diagnosed with lumbar spine stenosis and grade 1 spondylolisthesis;[108] and he was recommended for spinal surgery evaluation. The medical examiner, a doctor of chiropractic, did not note that he found this condition during the physical examination or that the driver had a limp or other orthopedic, muscular, or neuromuscular disease that could interfere with his ability to safely control and operate a CMV.

According to the FMCSA, an examiner must evaluate whether the driver has a perceptible limp; sufficient mobility and strength of the spine or torso to safely drive and perform other job tasks; limitations of motion of the spine or torso; and spine, torso, or other musculoskeletal tenderness. The examiner is required to determine if the severity of a reversible or progressive musculoskeletal disease interferes with driving ability, the likelihood of progressive limitation, and the potential for gradual or sudden incapacitation. If the findings from the examiner dictate, radiology or other examinations should be used to diagnose conditions such as spondylolisthesis.[109] It is unknown whether the school bus driver's increasing lower back, right hip, and leg pain—a result of his lumbar spine stenosis and grade 1 spondylolisthesis—may have contributed to the crash. However, the driver was experiencing enough difficulty walking that his orthopedic doctor had noted this symptom in his medical records and increased his prescription medications to include both tramadol and oxycodone, a narcotic drug.

Nervous or Psychiatric Disorders. The school bus driver reported that he had mild depression and was taking medications for depression and anxiety. According to FMCSA guidance, the medical examiner has a fundamental obligation during the psychological assessment to establish whether an applicant has a psychological disease or disorder that increases the risk for periodic, residual, or insidious onset of cognitive, behavioral, or functional impairment, because: (1) a CDL driver with a mood disorder may, during a depressive episode, exhibit slowed reaction time and poor judgment; and (2) some personality disorders can directly affect memory, reasoning, attention, and judgment, and a CDL driver with a mood disorder may,

---

[108] Spondylolisthesis is a condition in which the vertebra—most commonly the fifth lumbar vertebra—slips forward. Patients complain of low back pain as well as sciatica if there is nerve root compression. Medical treatment consists of bed rest, bracing, and analgesics; if symptoms are unremitting or frequent and severe enough to require the regular use of analgesics, various surgical procedures including spinal fusion are available. In aviation, flight certification based on analgesics is unadvisable, and a waiver request would be in order four to six months postsurgery to ensure full healing (Gaines and Humphrey 1993; Rayman and others 2001).

[109] See nrcme fmcsa.dot.gov/mehandbook/muscul_system_ep.aspx, accessed May 20, 2013.

during a depressive episode, exhibit slowed reaction time and poor judgment.[110] Further, personality disorders, depending on severity and type, "may exhibit inflexible and maladaptive behaviors and have an increased crash rate." The FMCSA guidance also states that risk factors associated with personality disorders can interfere with driving ability by compromising:

- Attention, concentration, or memory affecting information processing and the ability to remain vigilant to the surrounding traffic and environment.

- Visual-spatial function (for example, motor response latency).

- Impulse control, including risk taking.

- Judgment, including the ability to predict and anticipate.

- Ability to problem solve, including response to simultaneous stimuli in a changing environment and potentially dangerous situations.

In a review of the medical examiner's evaluation of the school bus driver, there was no indication that the depression and anxiety reported by the driver and treated with medications were appropriately evaluated and forwarded for further followup.

Frequent Alcohol Use. The school bus driver checked "yes" on his CDL examination form, under the health history section, that he engaged in "regular, frequent use of alcohol" and wrote that he had "2 glasses of wine per day." The driver did not self-report his family doctor's diagnosis of alcoholism in October 2011; his use of prescription medicines to treat the symptoms of alcoholism; or the diagnosis of alcohol abuse in November and December 2011, including the doctor's advice to abstain and seek therapy, which would be considered the additional information or limitations required by the FMCSA. In his postcrash interview, the driver reported his alcohol ingestion (he reported drinking two "double" scotches), which did not match what he reported to the CDL medical examiner. According to the DOT, medical examiners do not give applicants drug or alcohol tests and largely base determinations on the physical examination and driver identification of prescribed medications, alcohol use, or illegal drug use (GAO 2012).

Except with a current clinical diagnosis of alcoholism, the medical examiner makes the final determination as to whether the applicant meets FMCSA medical standards for driver certification. However, there is no question on the CDL form to report a current clinical diagnosis of alcoholism, which would provide a starting point from which an examiner could make a CDL qualification determination. The FMCSA guidance is to use whatever tools or additional assessments the examiner feels are necessary;[111] the *Medical Examiner Handbook* states that the examiner may use standardized drug or alcohol abuse screening tests.[112] If the applicant shows signs of alcoholism, the examiner is advised to have the driver consult a

---

[110] See nrcme.fmcsa.dot.gov/mehandbook/psych4_ep.aspx#, accessed May 16, 2013.

[111] See nrcme.fmcsa.dot.gov/documents/Complete%20Guide%20to%20ME%20Certification%20final%-20110112.pdf, accessed April 18, 2013.

[112] Screening tests include the Alcohol Use Disorders Identification Test (AUDIT), Cage, and T-ACE.

specialist for further evaluation. Despite the school bus driver reporting frequent alcohol use to the medical examiner, he was not forwarded for consultation with a specialist.

When the school bus driver received his medical certification in January 2012, medical examiners were not required to have any training and did not need to demonstrate any special competence to medically certify commercial drivers. The FMCSA had no regulatory authority over the examining doctor, there were no federal training and certification programs to ensure that examiners were familiar with the regulations, and there was no national registry of examiners. However, examiners were expected to exercise good medical judgment and carefully evaluate each person on whom they performed a physical examination (Herner, Smedvy, and Ysander 1966).

The medical examiner must follow the medical standard in 49 CFR 391.41 and also has access to FMCSA guidelines, which are intended as standards of practice and are based on the medical literature.[113] The FMCSA provides a website containing medical advisory criteria and posts seminar reports. Although the medical examiner must rely on the driver applicant to accurately and completely list his or her medications, the examiner is responsible for determining medical qualifications under the FMCSRs and, as such, may rely on reporting by the driver's prescribing physician on medication effect (Kreuger and Leaman 2011).

In this case, the medical examiner, a doctor of chiropractic, had never seen the school bus driver before and did not obtain medical records or consults from the driver's primary care and specialist doctors. The extent to which the medical examiner verbally discussed the driver's chronic low back pain and diagnosis of alcoholism (and alcohol abuse), which are underlying potentially disqualifying conditions, is not known. Moreover, it is unknown to what extent he discussed the prescription medications he was aware of, even though he wrote on the CDL form that they do not interfere with driving. Therefore, the NTSB concludes that the CDL medical examiner did not thoroughly evaluate the school bus driver for medical conditions that could have disqualified him from receiving a CDL. The NTSB also concludes that based on the school bus driver's combination of medical conditions and use of multiple prescription medications, it is likely that he would not have been medically certified to drive a school bus if: (1) he had fully disclosed his medical history on the CDL medical certification examination form, or (2) the medical examiner had completed a more thorough evaluation.

2.3.4.3 Anticipated Improvements in CDL Medical Examination Process. The Chesterfield crash investigation is not the first in which the NTSB has been concerned with the FMCSA medical examiner process, the certification of drivers with potentially disqualifying medical conditions and prescription medications, and the training and qualifications of medical examiners. In 2001, the NTSB issued numerous recommendations that called for a systemwide modification to the medical oversight of commercial drivers following a medical impairment-related motorcoach crash that occurred in New Orleans, Louisiana (NTSB 2001b). Included in these recommendations to the FMCSA were the development of a comprehensive medical oversight program for interstate commercial drivers that includes: (1) qualifying medical examiners and training them on driver occupational issues; (2) periodically updating medical

---

[113] If the medical examiner chooses not to follow the guidelines, the reason(s) for the variation should be documented.

certification regulations; (3) providing specific guidance and a readily identifiable source of information for those performing such examinations; and (4) preventing, or identifying and correcting, the inappropriate issuance of medical certification. Of these four program elements, Safety Recommendations H-01-17, -19, and -20, respectively, are currently classified "Open— Acceptable Response," and Safety Recommendation H-01-21 is classified "Open—Unacceptable Response."

Since the NTSB made these recommendations, the FMCSA has created a Medical Division and continues to work toward an effective commercial driver medical oversight system, including publication of the *Medical Examiner Handbook.*

New federal regulations will go into effect in May 2014 requiring CDL holders to obtain their medical examinations from trained and certified physicians. In April 2012, the FMCSA published a final rule establishing a National Registry of Certified Medical Examiners.[114] The rule requires all healthcare professionals responsible for issuing medical certificates for interstate truck and bus drivers to complete a training course and to pass an examination to assess their ability to apply the rules and advisory criteria. In May 2013, the FMCSA published an NPRM proposing a new medical examination form and revisions to the system used by medical examiners to report the results of all completed commercial driver physical examinations to the FMCSA and state driver licensing agencies.

Prior to 1992, only medical doctors and doctors of osteopathy were allowed to perform CDL medical examinations.[115] Title 49 CFR 390.5 modified the regulations to allow certain state-licensed, -certified, or -registered health care professionals to perform examinations, including medical doctors, osteopaths, physician assistants, advanced practice nurses, and doctors of chiropractic.

The federal regulations subject to compliance beginning May 21, 2014, will still allow the types of medical professionals listed above to issue medical certificates as long as they complete a training course, pass an examination test, and become listed on the National Registry of Certified Medical Examiners. In 2001, the NTSB expressed concern about the inability of most doctors of chiropractic—who may serve as examiners in many states—to assess the possible effects of prescription drugs, nonprescription drugs, and drug interactions on commercial vehicle operators (NTSB 2001b). The practice of chiropractic medicine specifically avoids the use of medications; doctors of chiropractic typically receive no formal training in pharmacology and are not licensed to prescribe medication. It is, therefore, unreasonable to expect that most doctors of chiropractic would know how certain medications affect driving performance. Similarly, other professionals who are permitted to perform physical examinations in the state in which they are licensed may not be permitted to prescribe medications, and thus would also have inappropriate backgrounds to make such determinations.

---

[114] See 77 FR 24104, April 20, 2012.

[115] Medical certification, which qualifies an individual as being fit to drive a commercial vehicle, became a federal requirement under the Motor Carrier Act of 1935. The first physical qualification standards were established in 1939; included in the requirements was that the driver be in good physical and mental health and have no addiction to alcohol or narcotic drugs. See 57 FR 33276, July 29, 1992.

Because the medical examiner remains the certifying authority, he or she must have sufficient background to analyze and evaluate medication effects. A brief training program for examiners cannot replace formal courses in pharmacology and current prescribing experience to the extent necessary for such evaluations. The NTSB remains concerned that the FMCSA is choosing to permit those whose backgrounds would not meet minimum requirements to make appropriate certification decisions to apply for inclusion on a registry of qualified examiners. The NTSB concludes that some medical professionals who are authorized to perform medical examinations and certify commercial drivers as fit to drive may lack the knowledge and information critical to certification decisions; consequently, drivers with serious medical conditions may not be sufficiently evaluated to determine whether their conditions pose a risk to highway safety. The NTSB recommends that the FMCSA require that all persons applying for inclusion on the National Registry of Certified Medical Examiners have both a thorough knowledge of pharmacology and current prescribing authority.

Despite the numerous anticipated improvements in the CDL medical examination process expected with the recent rulemaking by the FMCSA, the NTSB maintains that more action is necessary to ensure a comprehensive medical oversight program. In a review of the circumstances of this crash and the issuance of medical certification to the school bus driver, NTSB investigators determined that the medical examiner did not have all of the information and guidance needed to make an informed decision regarding the applicant's various medical conditions and prescription drugs. Furthermore, the system did not include a means to review the medical examiner evaluation and correct the possible inappropriate issuance of the certification. Safety Recommendations H-01-17, -19, -20, and -21, from the New Orleans crash investigation, addressed these concerns 12 years ago, and they remain in an open status. Therefore, the NTSB reiterates Safety Recommendations H-01-17, -19, -20, and -21 to the FMCSA.

2.4 Mack Roll-Off Truck

2.4.1 Postcrash Brake and Overweight Conditions

The truck was equipped with four axles, corresponding to eight individual brake assemblies. Under CVSA OOS criteria, a vehicle should be placed out of service when the number of defective brakes is 20 percent or more of the total number of brakes on the vehicle. The NTSB postcrash inspection found that four of the eight truck brakes (50 percent) would have been considered defective. The pushrod stroke on the left side of axle 4 exceeded the adjustment limit by 0.25 inch, which would be considered a defective brake under CVSA OOS criteria. However, the other three OOS defects were hidden from view by other vehicle components and would not have been identified during a standard roadside inspection.[116] Therefore, had the truck been subjected to a roadside inspection on the day of the crash, only one of the eight brakes would have been identifiable as defective, and the vehicle would not have been placed out of service for defective brakes.

---

[116] The void in the edge of the brake pad lining on the left side of axle 2, the loose brake pad lining on the right side of axle 3, and the oil and grease contamination on the right side of axle 4 would all be considered defective brake lining conditions—and, therefore, defective brakes under CVSA OOS criteria. These defects would not likely have been detectable without removing the wheels from the vehicle and thus would not have been detected during either a routine pretrip or roadside inspection.

Because four defective brakes were found postcrash, the NTSB conducted brake efficiency calculations to determine what effect, if any, the condition of the service brakes would have had on the ability of the truck to stop in an emergency situation, such as to avoid striking the school bus. In addition, because several overweight conditions were identified on the truck postcrash, the NTSB also considered vehicle weight in assessing the truck's braking efficiency (Heusser 1991):[117]

- At 84,950 pounds, the truck was found to exceed the legal maximum registration weight of 80,000 pounds.

- Additionally, the GAWR of 24,680 pounds for both drive axles was exceeded; the forward drive axle (axle 3) had a measured weight of 29,000 pounds, and the rear drive axle (axle 4) had a measured weight of 28,050 pounds.[118]

- Under New Jersey state law, when considered as a group, the three rear axles would not be allowed to carry more than 56,400 pounds. The total weight being carried by this group (axles 2, 3, and 4) when measured at the crash scene was 69,850 pounds— or 13,450 pounds over the maximum legal axle group weight.

Based on brake force efficiency calculations, the braking ability of the truck was reduced by approximately 9 percent when compared to the same vehicle with a legal weight,[119] as shown in table 6.

Table 6. Reduction in braking efficiency of truck due to weight.

Weight	Brakes	Deceleration Rate (g)	% Change
Legal weight	Accident brakes	0.360	- 9.2
Accident weight	Accident brakes	0.327	

---

[117] A vehicle's braking efficiency is a good measure of its ability to stop on a given surface. Air brake efficiency is determined by comparing a vehicle's *attempted* braking force to its *available* braking force, and is expressed as a percentage of how the calculated brake performance compares to ideal brake performance (100 percent brake efficiency). When available braking exceeds attempted braking, only a portion of the available friction is used, and the overall braking efficiency of the vehicle is reduced. Air brake efficiency calculations were conducted using the Heusser method in combination with the information gathered from the truck during the inspection and subsequent investigation.

[118] The weight being carried by axles 3 and 4 was 13.5–17.5 percent greater than that designated by the manufacturer. Although the GAWRs for the drive axles had been exceeded, this condition, in and of itself, did not adversely affect the handling of the truck during the collision sequence.

[119] Legal weight was calculated by taking the lesser of the following values: GAWR for the axle, tire weight ratings for the tires on the axles, or the maximum allowable weight under New Jersey state law for an axle or axle group. The legal weight distribution for the truck was determined to be 18,000 pounds on axle 1 (based on GAWR), 18,740 pounds on axle 2 (based on tire weight ratings), and 18,830 pounds on each of axles 3 and 4 (based on New Jersey state law for the axle group), for a total legal weight of 74,400 pounds.

The overweight condition of the truck reduced its deceleration rate by approximately 9 percent, while the condition of the brakes accounted for a reduction of approximately 17 percent. (See table 7.) Brake efficiency calculations revealed that the truck was found to have a 52–67 percent range of braking efficiency values. This range was affected by the overall weight of the truck; the condition of its brakes; and the combination of its weight and brake conditions, which reduced the overall potential deceleration by approximately 32 percent. (See table 8.)[120]

Table 7. Reduction in braking efficiency of truck due to condition of brakes.

Weight	Brakes	Deceleration Rate (g)	% Change
Accident weight	Legal brakes	0.397	- 17.6
Accident weight	Accident brakes	0.327	

Table 8. Reduction in braking efficiency of truck due to weight and brakes combined.

Weight	Brakes	Deceleration Rate (g)	% Change
Legal weight	Legal brakes	0.486	- 32.7%
Accident weight	Accident brakes	0.327	

The NTSB concludes that the combination of the truck's defective brakes and overweight condition reduced its overall braking efficiency, thereby contributing to the severity of the crash.

2.4.2 Vehicle Weighing Systems

Accurate vehicle weight is difficult to determine without the use of a scale. This task becomes even more difficult when the vehicle is loaded in the field, as applies to aggregates or earthen construction materials, raw natural resources, and garbage or refuse, or logging and timber operations, or agricultural operations. Current weight laws or enforcement practices in many states allow a vehicle to travel from the location where the load is picked up to the nearest publicly available scale facility. If the truck is overweight, the driver is supposed to return to the loading facility to reduce the load. As evident in this crash, vehicle weight can affect numerous vehicle systems; several overweight conditions were identified on the truck that affected its braking efficiency.

The loaded vehicle was not weighed at the construction site. To comply with New Jersey state law and the NJDOT prohibition on overweight vehicles transporting material on the roadways, the driver was required to go a weigh station 2.5 miles from the site. However, instead

---

[120] Potential variance in roadway friction coefficient values would also affect overall deceleration.

of stopping at the weigh station, he drove directly to the Herman's Trucking facility, which was 9 miles away, to weigh the bin container.

Onboard vehicle weighing system technology incorporates load cells, air pressure transducers, or strain gages to instantly identify vehicle weight. This technology can be used in the field, reducing the chances of overweight vehicles leaving a loading site and operating on public roads. Depending on the type and configuration of the vehicle, sensors are most commonly placed on the axles, the suspension components, or between the cargo body and vehicle frame. These systems include a driver interface that typically displays individual axle weights, axle group weights, cargo weight, and gross vehicle weight. It is usually mounted within the cab, either between the seats or in another location where it would not be distracting to the driver while the vehicle is in motion. After installation, the systems are programmed with the vehicle parameters and calibrated to the empty weight of the vehicle to allow for accurate reporting of gross vehicle weight.

Onboard vehicle weighing systems are estimated to cost $2,000–$7,500 for typical systems that are accurate to 0.5–3 percent.[121] Systems that are certified to be "legal for trade," which are accurate to better than 10 pounds,[122] can cost as much as $15,000.[123] Costs are dependent on the type of system, type of vehicle, and how it is purchased. Discounts are often applied to multiple systems installed within a fleet. Automated electronic trailer identification is available on many systems, which allows a truck to pull a variety of trailers, with the system automatically adjusting for the preprogrammed trailer characteristics.

The NTSB concludes that had the truck been equipped with an onboard weighing system, the truck driver would have had information about its overweight condition. Therefore, the NTSB recommends that the National Highway Traffic Safety Administration (NHTSA) develop minimum performance standards for onboard vehicle weighing systems for trucks that have a GVWR of 10,000 pounds or more and are typically field loaded and used in the transportation of aggregates or earthen construction materials, raw natural resources, and garbage or refuse, or in logging and timber operations, or agricultural operations. Once minimum performance standards for onboard vehicle weighing systems are established, the NTSB recommends that NHTSA require these systems to be installed on newly manufactured trucks that have a GVWR of 10,000 pounds or more and are typically field loaded and used in the transportation of aggregates or earthen construction materials, raw natural resources, and garbage or refuse, or in logging and timber operations, or agricultural operations.

---

[121] See www.vishaypg.com/onboard-weighing/onboard-weighing/, www.ricelake.com/products/on-board-weighing, www.kanawhascales.com/OBW/index.php, and www.transdata.us/, accessed January 11, 2013.

[122] Vishay Precision Group reported that its onboard weighing systems are accurate to 1–3 percent (depending on the use of load cells or air ride pressure sensors) and cost $2,000–$5,000. Rice Lake Weighing Systems reported that its standard systems are accurate to approximately 1 percent and cost $3,000–$7,500. Kanawha Scales & Systems reported that its systems are accurate to 0.5–1 percent and cost approximately $6,000 installed on a triaxle dump. Trans-Data On-Board Weight Systems reported that its systems are accurate to 1–1.5 percent and cost $4,500–$5,000.

[123] Rice Lake Weighing Systems reported that its "legal for trade" system is accurate to better than 10 pounds and typically costs $11,000–$15,000.

2.4.3 Onboard Brake Stroke Monitoring Systems

Although only one of the brakes on the truck was found to be out of adjustment due to pushrod stroke, two were at the adjustment limit, and one was within 1/8 inch of the adjustment limit. Brake adjustment is difficult to gauge from inside the cab of a truck, and drivers do not typically get under their vehicles to check brake adjustment during a routine pretrip inspection. Doing so would require another person to assist in the application and release of the brakes, while the driver measured the pushrod stroke. For these reasons, it is possible that the truck driver was unaware of the condition of the brakes on his vehicle.

According to the FMCSA, brake stroke monitoring systems can provide valuable information to the driver to help maintain the vehicle's safe operation, and they can also aid motor carriers in identifying air brake adjustment and maintenance problems.[124] These systems can be included on a vehicle as original equipment or can be installed aftermarket on an older truck. Monitoring systems incorporate sensors either in the chambers or adjacent to the pushrods of pneumatically braked commercial vehicles. The sensors monitor the travel of individual pushrods to instantly identify wheel-specific, out-of-adjustment, nonfunctioning, or dragging brake issues. Some display units use tricolored light-emitting diodes to indicate types of brake faults and wheel locations. One such system displays a green light indicating that the brakes are in adjustment, a yellow light indicating that they are within 1/8 inch of adjustment, and a red light indicating that they are at or exceed the maximum allowable stroke adjustment limit. In the case of this crash, one yellow indicator light and three red indicator lights would have been illuminated on a display unit of this type.

As described in section 1.6.3 and discussed in section 2.4.5 below, Herman's Trucking periodically made manual brake slack adjustments to its trucks. In a recent NTSB crash investigation involving a truck-tractor semitrailer that struck a passenger train in Miriam, Nevada, it was found that the motor carrier had also been manually adjusting the vehicle's automatic slack adjusters (NTSB 2012). Although the FMCSA has developed a product guide for onboard brake stroke monitoring systems, there are currently no standards or requirements for the systems on air-braked commercial trucks, such as the Miriam and Chesterfield trucks.

The NTSB concluded in the Miriam investigation that had the truck been equipped with an onboard brake stroke monitoring system, the truck driver would have had advance information about the out-of-adjustment and inoperative brakes. The NTSB issued Safety Recommendations H-12-58 and -59 to NHTSA to develop minimum performance standards for onboard brake stroke monitoring systems for all air-braked commercial vehicles; and to then require that all newly manufactured air-braked commercial vehicles be equipped with those systems. The NTSB concludes that had the Chesterfield truck been equipped with an onboard brake stroke monitoring system, the truck driver could have had information about the out-of-adjustment and impending out-of-adjustment brakes. Therefore, the NTSB reiterates Safety Recommendations H-12-58 and -59 to NHTSA.

---

[124] See www.fmcsa.dot.gov/facts-research/systems-technology/product-guides/brake-stroke.htm, accessed May 16, 2013.

2.4.4 Vehicle Lift Axle Air and Brake System Installation

The original air system installed on the chassis by Mack Trucks was later modified when the final stage manufacturer, AWE, installed the roll-off hoist components and lift axle. Mack Trucks had made available to all intermediate and final stage manufacturers the *Body Installer's Guide for Mack Class 8 Chassis* (Mack 2003). However, AWE's installation was inconsistent with the guide's written instructions for the lift axle. The guide stated that for the vehicle to comply with FMVSS 121, the liftable axle should have its own service brake relay valve (that is, a second relay valve should have been added to the system to supply air directly to the lift axle brakes). AWE added only a single quick-release valve to service the two brake chambers.[125] The service brake relay valve was thus supplying air to six brake chambers rather than four brake chambers as designed. The additional lift axle service brake relay valve was not added to the system on the Mack roll-off truck. Figure 21 shows, within shaded blue lines, the components that AWE added to the truck and the method for installing the lift axle air brake piping. The partial air system schematic, presented in figure 22, shows the Mack Truck-recommended brake air piping for a lift axle (enclosed within shaded pink lines).[126]

The AWE installation process resulted in improper piping in which the air brake lines forced greater air volumes through the same size valves (air for six brake chambers through the relay valve designed for four, and air for four brake chambers through the modulator valve designed for two), thus increasing the time required to supply the necessary volume and pressure of air to the brake chambers. In 2004, when the truck was manufactured, FMVSS 121 required for new vehicles that the air pressure in each brake chamber be capable of reaching 60 psi within 0.45 second and also be capable of dropping from 95 to 5 psi within 0.55 second. The NTSB testing revealed that several of the truck's brake chambers did not meet these requirements.

A PARO representative stated that the lift axle supplier trained its staff to install the lift axle brake piping. However, its mechanics have since modified installation procedures and now install relay valves with the lift axles based on information provided by another lift axle supplier and technical bulletins from the National Truck Equipment Association (NTEA), of which PARO is a member.

---

[125] AWE added a "T" type connection to the output side of the left drive axle ABS modulator valve, a section of piping between the modulator valve and the quick-release valve, and two additional sections of piping from the quick-release valve to each of the two brake chambers. This installation method created a situation where the one ABS modulator valve, designed to regulate (service) two brake chambers, was now regulating (servicing) four brake chambers.

[126] This schematic represents the easiest installation procedure, in which there is no ABS modulation of the lift axle brakes. If ABS modulation of the lift axle brakes is preferred, the *Body Installer's Guide for Class 8 Chassis* presents other brake air piping options.

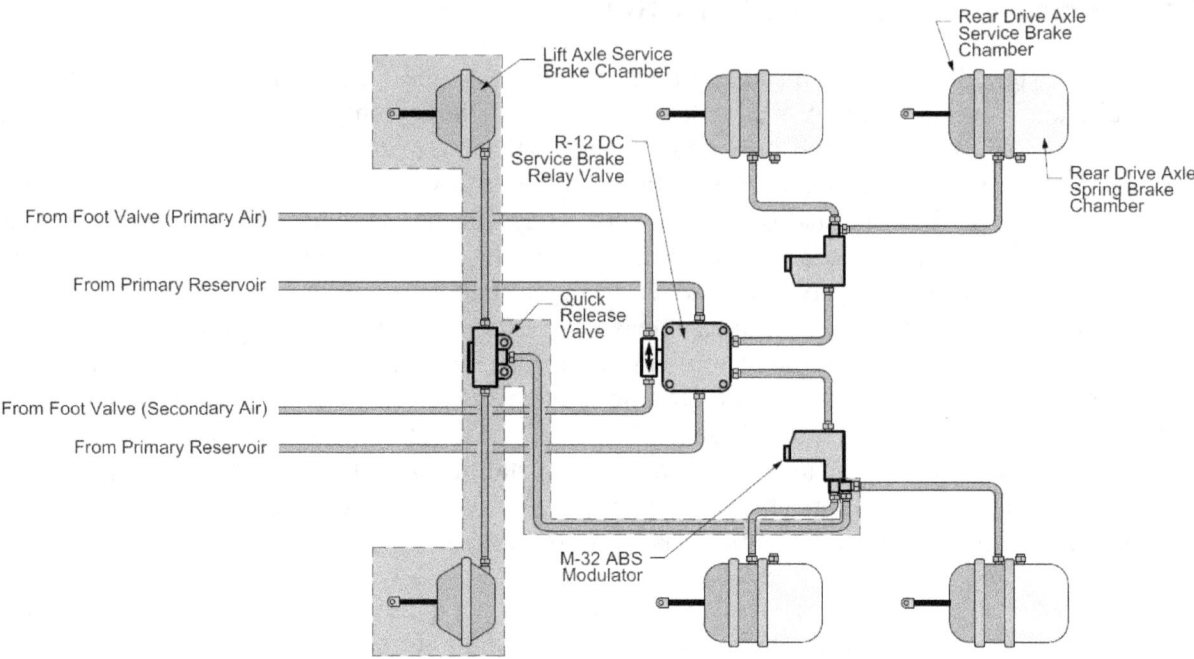

Figure 21. Partial diagram of air system showing actual piping of lift axle brake components on truck, outlined and shaded in blue.

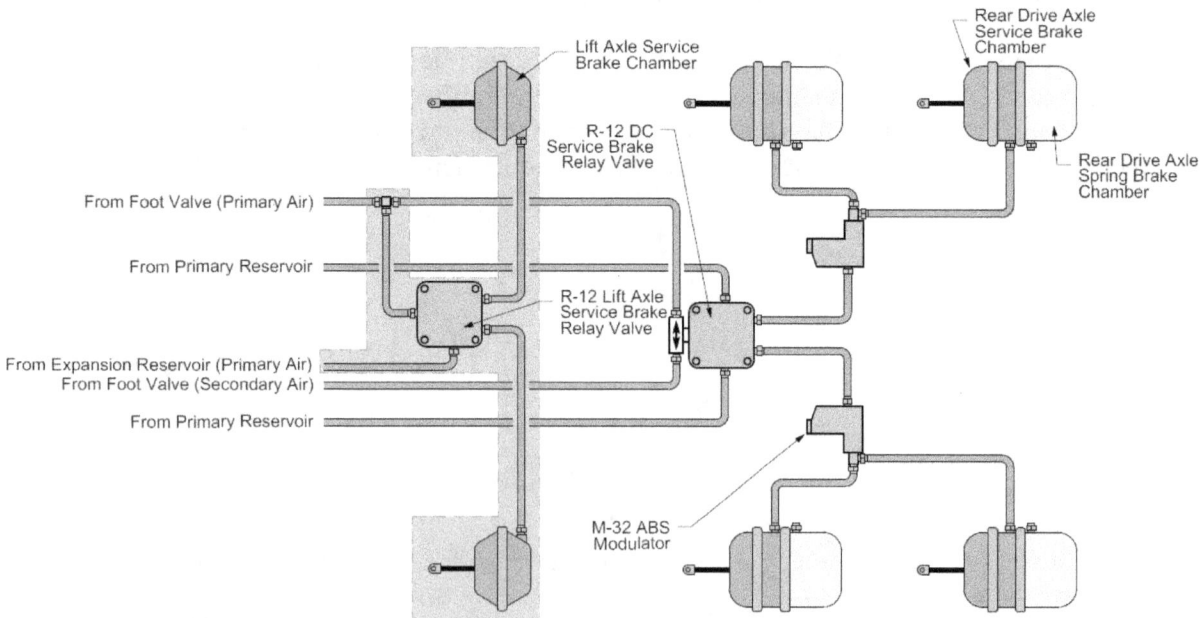

Figure 22. Partial diagram of Mack Truck-recommended piping of lift axle brake component air system, outlined and shaded in pink.

Although strict adherence to the guidance instructions provided by Mack Trucks was not required, the installation methods used by AWE as the final stage manufacturer were required to meet FMVSS 121 when the vehicle was sold. It was the responsibility of AWE to certify that its installation method met FMVSS 121. The Mack installation guide cautions on required compliance with all applicable federal standards. Specifically, in the "liftable axle brake air piping" section, it states: "it is the responsibility of the body equipment installer or alterer to ensure that the vehicle remains in compliance with U.S. FMVSS 121 when modifications are made to the air brake system"[127] (Mack 2003).

Postcrash, PARO acknowledged that at the time the truck was manufactured (in 2004, by AWE), its mechanics did not follow the Mack-recommended plumbing and installation guidance for the lift axle. Moreover, PARO could not provide documentation to the NTSB that its installation method was tested to meet and comply with FMVSS 121 prior to self-certification.[128] Based on the results of postcrash testing of the truck and the lack of actual timing testing performed by the final stage manufacturer to show otherwise, the NTSB concludes that due to improper installation of the lift axle brake system by AWE, the brakes on the truck would not likely have met FMVSS 121 timing requirements at the time of manufacture. The NTSB further concludes that had the lift axle brake system been properly installed on the truck, air would have been applied to the brakes earlier, thereby reducing the severity of the crash. Since the manufacture of the truck in 2004, PARO reports that it has modified its installation procedures based on NTEA technical bulletins. Therefore, the NTSB recommends that the NTEA notify its members of the Chesterfield crash and of the need to check their vehicles for potentially improper lift axle brake installation.

2.4.5 Herman's Trucking Oversight

Herman's Trucking is an authorized for-hire and private carrier of building materials, general freight, machinery, and construction and landscaping materials. At the time of the crash, the company had a fleet of 23 drivers and 21 vehicles. According to the FMCSA, the company was subject to 25 roadside inspections in the 24 months preceding the crash. During this time, 14 vehicle inspections were conducted—with a 14 percent OOS rate (two OOS violations), which is below the national average OOS rate of 20.7 percent. As of February 16, 2012, Herman's Trucking had not exceeded the thresholds for any of the seven BASIC safety measurement system values. The FMCSA had conducted compliance reviews on Herman's Trucking in March 2006, January 2001, and October 1996—each of which resulted in a satisfactory rating.

Postcrash, the FMCSA (through the NJSP) conducted a review and inspected 17 vehicles, which resulted in one OOS violation (a defective taillight). In contrast to the limited violations found during inspection of the rest of the Herman's Trucking fleet, the NTSB postcrash

---

[127] To examine the extent to which this incorrect installation process may apply to other roll-off trucks, the NTSB inspected 19 trucks with lift axles installed by various final stage manufacturers, two of which were completed by PARO. Dedicated lift axle service brake relay valves were present on all of the vehicles inspected.

[128] During an NTSB investigative meeting with PARO, its general manager was asked whether any 121 or other FMVSS testing was performed, and he stated that it was not. The only additional testing AWE had done was in reference to some weight or weight distribution requirements for a particular state.

inspection of the truck found that four of the eight truck brakes (50 percent) would have been considered defective.

Herman's Trucking personnel stated to NTSB investigators that its vehicles were inspected biweekly and that manual brake slack adjustments were performed if necessary. If an out-of-adjustment brake was equipped with an automatic slack adjuster, mechanics would adjust the slack adjuster one time and mark the vehicle frame with a grease pencil to indicate such. If that brake was found to be out of adjustment a second time, the slack adjuster would be replaced. The NTSB has previously issued recommendations advising against manual adjustment of automatic slack adjusters (NTSB 2006). Although manual adjustment may temporarily bring the brake into compliance, it will not retain adjustment due to defective components or the slack adjuster itself, resulting in reduced braking ability until the root of the problem is addressed.

NTSB investigators also reviewed Herman's Trucking oversight of the transportation of loads from the construction site to the recycle location. The company was contracted to provide a roadway construction site with empty roll-off bin containers and to haul away and recycle the contents. To comply with New Jersey state laws, the driver of the truck was required to go to a weigh station located 2.5 miles from the site.[129] However, instead of stopping at the weigh station, drivers went directly to the Herman's Trucking facility and then had the bin container weighed. The NTSB concludes that the driver of the truck did not ensure that his vehicle was within allowable weight restrictions, despite the requirement to do so and the fact that there was a weigh station located 2.5 miles from his originating point.

Postcrash, the NTSB examined nine company weight slips and found that five showed the truck and load being over the 80,000-pound limit, which indicates that the company had no procedures or policies in place to ensure compliance with the restriction on overweight loads.[130] In March 2012, as a result of the crash, the NJSP issued citations to Herman's Trucking after a determination that this truck was being operated in an overweight condition. Accordingly, the NTSB concludes that Herman's Trucking did not have effective oversight of its drivers to ensure that the company truck was routed, as required by law, to the closest weighing scales, thereby preventing the transportation of overweight loads on public roads. The NTSB recommends that

---

[129] Per *New Jersey Statutes Annotated* 39:3-20, the fine for overweight vehicles is $523 plus $100 for each 1,000 pounds (over the specified weight) or fraction thereof.

[130] Offloading an overweight vehicle (to reduce load weight) at a scale to avoid a citation is not allowed. If a truck is found to be overweight by an enforcement officer at a scale, a citation is issued. The driver is required to then reduce the load (if divisible) or obtain an overweight permit (if not divisible) before continuing on any public road. A divisible load is any cargo that can be separated into a legal weight without affecting its integrity, such as aggregate (sand, topsoil, gravel, and stone), logs, scrap metal, fuel, milk, or trash/refuse/garbage. By contrast, if the load contains a singular piece or item that cannot be separated into units of lesser weight without affecting its physical integrity, it is considered to be a nondivisible load. In the case of the Chesterfield truck, the driver was permitted to depart the construction site, travel to the closest available scale (state-controlled or publicly available), and have the vehicle weighed. When a vehicle is found to be overweight, the driver needs to reduce, transfer, or redistribute the load so that the overweight vehicle does not operate on public roads (including being prohibited from returning to the construction site). Should an overweight vehicle be stopped by law enforcement prior to arriving at a scale, the driver will not be cited for being overweight provided that he or she can prove that they are on their way to the scale. However, once the vehicle driver travels past the closest scale or is not making an attempt to travel to it, law enforcement may issue a citation for weight-related violations.

Herman's Trucking develop and implement route procedures to prevent the transportation of overweight loads.

2.5 Highway Issues

This section considers the design of the BCR 528–660 intersection and how intersection conditions may be related to the cause of the crash, reviews intersection crashes in general, and assesses potential technological solutions.

2.5.1 Intersection Design

The BCR 528–660 intersection was skewed at an angle of 63 degrees, and the sight distance at the stop line was limited due to trees located alongside the roadway. The NTSB evaluated the potential role these conditions played in the cause of the crash. The stop line on BCR 660 was set back to accommodate left-turning trucks from BCR 528; and pine trees were located to the south of BCR 528 eastbound, approximately 23 feet from the travel lane.

Postcrash sight distance testing was conducted with an exemplar school bus positioned at eight separate locations to determine the ability of the school bus driver to see the approaching truck on BCR 528 (departure sight distance) and of the truck driver to see the stopped school bus (stopping sight distance). From the white stop line, it was determined that a driver on BCR 660 could see only 206 feet to the left. Likewise, the position of the stop line reduced the distance from which a vehicle on BCR 528—in this case, the truck—could see the potential cross traffic waiting on BCR 660. The truck driver could see a school bus at the stop line from 287 feet, which is below the required stopping sight distance of 360 feet.

Despite the apparent sight distance obstruction presented by the trees located west of the intersection and parallel to the BCR 528 eastbound travel lane, the school bus driver was adamant in his interview that he was aware of the conditions at the intersection and that he had pulled forward sufficiently before entering the intersection, such that he was able to see clearly in both directions. Additional sight distance testing determined that at the location where NTSB investigators believe that the bus driver came to a stop, the departure sight distance was 463 feet, and the sight distance for the truck driver was 686 feet. If the bus driver had stopped closer to the BCR 528 travel lane (within 10 feet), he would have had an available sight distance of 783 feet. The NTSB concludes that though the sight distance available for vehicles stopping at the stop line on northbound BCR 660 was inadequate, it did not contribute to the cause of the crash because the school bus driver stopped forward of the stop line, where there was sufficient sight distance.

The NTSB also evaluated the geometry of the BCR 528–660 intersection. A right-skewed intersection, such as the crash intersection, can affect a driver's line of sight as well as the extent to which a driver needs to turn to look to the left to see approaching traffic. The accident reconstruction determined that the truck was approximately 368–406 feet west of the AOI and at an angle of approximately 110 degrees to the left of the heading of the school bus. If the intersection was not skewed at 63 degrees and was closer to a right angle, the school bus driver would not have had to turn his head and upper body as far to the left to observe oncoming traffic.

Therefore, the NTSB concludes that the skew of the BCR 528–660 intersection required the school bus driver to turn farther to the left to observe oncoming traffic, which called for more continuous scanning to safely cross the highway.

The BCE, working with NTSB investigators, reviewed the geometry, sight distance, and operation of the BCR 528–660 intersection following the crash. As a result of the review, Burlington County took several actions to improve the safety of the intersection, including:

- Placement of larger (36-inch-square) STOP signs, with reflectorized posts, on BCR 660.

- Placement of 40-mph warning signs on the already posted intersection warning signs on BCR 528, east and west.

- Movement and replanting of 14 trees from the southwest quadrant of the intersection to improve the line of sight. Two trees were also removed from the northwest quadrant of the intersection.

Based on these improvements, the NTSB concludes that Burlington County took direct steps following the crash to improve the safety of the BCR 528–660 intersection.

2.5.2 Intersection Safety

The NTSB analyzed NHTSA Fatality Analysis Reporting System (FARS) and National Automotive Sampling System (NASS) General Estimates System (GES) data from 2002 through 2011. According to FARS, in 2011, 6,817 fatal crashes occurred at intersections, resulting in a total of 7,265 fatalities.[131] Although intersection fatal crash counts and fatalities declined gradually during the 10-year period, they represented 22 percent of all fatal crashes and fatalities on US public roads. From 2002 through 2011, an estimated 12.3 million people were injured in intersection crashes in the United States,[132] representing 49 percent of all people injured on public roads. Of the 80,005 fatal crashes occurring at intersections, 57,542 occurred at an intersection where at least one traffic control device was present (72 percent). (See table 9.) Crashes at intersections account for more than 20 percent of all fatalities on public highways, and new effective countermeasures are needed.

---

[131] A combination of vehicle- and crash-level variables (related to junction and intersection type) in the FARS database were used to identify intersection fatal crashes.

[132] GES data come from a nationally representative sample of police-reported motor vehicle crashes of all types, from minor to fatal. The information is used to estimate how many crashes occur and what happens when they occur; it was created to identify traffic safety problem areas, provide a basis for regulatory and consumer initiatives, and form the basis for cost and benefit analyses of traffic safety initiatives. See www.nhtsa.gov/NASS, accessed March 25, 2013.

Table 9. Fatal crashes at intersections by year, 2002-2011.

Year	Fatal Crashes (FARS)				Injury Crashes (GES)	
	No. of Crashes	Percent of All US Crashes	No. of Fatalities	Percent of All US Fatalities	No. of Injured	Percent of All US Injured
2002	8,876	23	9,730	23	1,483,558	51
2003	8,808	23	9,669	23	1,395,975	48
2004	8,679	23	9,451	22	1,333,554	48
2005	8,715	22	9,519	22	1,322,963	49
2006	8,340	22	9,105	21	1,238,002	48
2007	8,252	22	8,913	22	1,181,498	47
2008	7,461	22	8,063	22	1,101,610	47
2009	6,982	23	7,561	22	1,080,256	49
2010	7,122	24	7,710	23	1,117,578	50
2011	6,817	23	7,265	22	1,081,655	49
Total	80,052	22	86,986	22	12,336,650	49

The NTSB has long advocated the use of technology to avoid collisions, such as forward collision warning, adaptive cruise control, lane departure warning, and electronic stability control. Forward collision warning and adaptive cruise control offer solutions to collisions caused by vehicles running into the rear of stopped or slower moving vehicles. Lane departure warning and electronic stability control can help prevent run-off-the-road crashes. These types of collision avoidance technologies are currently on the NTSB Most Wanted List of critical changes needed to reduce transportation crashes and save lives. Section 2.5.3 examines technology that can assist in preventing intersection crashes.

2.5.3 Intelligent Transportation Systems Technology

Two major programs that have the potential to reduce intersection crashes are currently being developed under the DOT Intelligent Transportation System Strategic Plan (FHWA-JPO-12-019). These connected vehicle technology programs are the V2V and V2I Communications for Safety Initiatives.

Both programs are built on dedicated short-range communication (DSRC) radios sending and receiving information between vehicles and traffic devices. For the V2V portion of the system to work, all vehicles would have a DSRC radio connected to a GPS device. The radio transmits vehicle positions over the last several seconds to allow the cars receiving the information to predict the path and speed of surrounding vehicles. The range of the system is currently about 300 meters. The system interface warns a driver if one of the surrounding vehicles is expected to encroach on the driver's projected path. The V2I systems communicate between approaching vehicles and an intersection to provide information such as whether the gap between vehicles on a cross road is sufficient for safe crossing. It is anticipated that V2V communications will be able to assist in preventing 76 percent of crashes, thereby reducing fatalities and injuries.

The V2V and V2I technologies are currently undergoing field testing under the safety pilot program—a two-phase joint research initiative led by NHTSA and the Research and Innovative Technology Administration to examine connected vehicle technology and real-world applications. Phase one was driver acceptance clinics (DAC), and phase two is the currently ongoing model deployment (MD).[133] The DOT and its research partners conducted six light vehicle DACs from August 2011 to January 2012.[134] Sixteen V2V-equipped vehicles were used—two from each participating original equipment manufacturer.[135] Participants were ordinary drivers in controlled roadway situations,[136] such as test tracks and parking facilities. Based on research from the DACs, NHTSA has concluded that connected vehicle systems have the potential to address unimpaired driver crashes.

Separate DACs are to be conducted for trucks. The drivers will be commercial vehicle drivers operating in a safe, highly controlled closed course environment. The objective of these DACs is to collecting subjective driver acceptance data on integrated safety systems and driver vehicle interfaces. At least one truck DAC will be held in conjunction with a trucking industry event or a similar fleet/driver-focused show.[137]

The MD phase is based in Ann Arbor, Michigan. This major research initiative involves several modes within the DOT, several vehicle manufacturers, public agencies, and academia. It began in fall 2012 and is expected to last through fall 2013. Two of the key MD elements are the use of 75 miles of instrumented roadway (through 27 roadside units) and about 3,000 vehicles (cars, trucks,[138] and buses[139] using integrated, aftermarket, and retrofitted technologies).

---

[133] See ConnectedVehicleTechnologyFactSheet-081012[1]pdf, accessed May 22, 2013.

[134] The dates and locations were: August 2011 in Brooklyn, Michigan; September 2011 in Brainerd, Minnesota; October 2011 in Orlando, Florida; November 2011 in Blacksburg, Virginia; December 2011 in Fort Worth, Texas; and January 2012 in Alameda, California. See www.its.dot.gov/presentations/trb_2013/Safety_-Pilot_TRB_files/frame.htm, accessed May 22, 2013.

[135] A car consortium led by the Crash Avoidance Metrics Partnership (CAMP) is responsible for all aspects of the driver clinics, ranging from project management to testing, recruitment of potential drivers, and overall management. CAMP Vehicle Safety Communications 3 is a cooperative research organization made up of eight of the leading car manufacturers—Ford, General Motors, Honda, Hyundai-Kia, Mercedes-Benz, Nissan, Toyota Motor Engineering & Manufacturing North America, Inc., and Volkswagen Group of America.

[136] See www.its.dot.gov/safety_pilot/index.htm, accessed May 22, 2013.

[137] See www.its.dot.gov/research/safety_pilot_overview.htm, accessed May 22, 2013.

[138] Three trucks integrated with wireless crash warning devices will be part of the separate truck DACs.

[139] Because transit drivers are already highly trained on advanced equipment (for which much data on human factors have already been collected), transit use of V2V and V2I technologies will be modeled and simulated to explore the dynamics of a variety of potential incidents—to include buses, light rail, pedestrians, transit signal priority, interaction with other vehicle and traffic elements, and transit operations.

The MD phase is expected to span one year of data collection to validate the potential benefits of V2V technologies.[140] NHTSA analyses show that DSRC-based connected vehicle technology could address approximately 80 percent of the crash scenarios involving nonimpaired drivers.[141] Data from the DACs have shown that this technology could help prevent the majority of typical crashes, such as those at intersections.

The DOT continues to perform V2V research and testing to address more complex crash scenarios, such as in the case of Chesterfield, which would require intersection collision avoidance. NHTSA has committed to an agency decision in 2013 regarding whether this safety technology is sufficiently developed to support rulemaking for passenger vehicles;[142] however, a decision on commercial vehicles is not expected until 2014.

In the Chesterfield crash, the school bus driver failed to see the truck approaching on BCR 528 from the west. The presence of connected vehicle technology (including V2V) on both the truck and the school bus would have resulted in the bus driver receiving a continuous warning as he began to cross the intersection. Therefore, the NTSB concludes that connected vehicle technology could have provided active warnings to the school bus driver of the approaching truck and possibly prevented the crash. As a result, the NTSB recommends that NHTSA develop minimum performance standards for connected vehicle technology for all highway vehicles; and, once the minimum performance standards are developed, require this technology to be installed on all newly manufactured highway vehicles.

On May 28, 2013, the NTSB submitted comments to the Federal Communications Commission NPRM to revise Part 15 of its rules to permit operation of unlicensed national information infrastructure (U-NII) devices within the 5-gigahertz (GHz) band.[143] Based on those comments, the National Telecommunications and Information Administration (NTIA) is beginning its evaluation process to test the use of U-NII devices on the 5-GHz band. A recently released report identifies a number of risk elements associated with the likelihood of harmful interference from large numbers of U-NII devices and concludes that further analysis is required to determine how to mitigate the identified risk factors (NTIA 2013).

The 5-GHz band—specifically, the frequency band between 5.850 and 5.925 GHz— serves as the platform for connected vehicle technologies essential to the advancement of transportation safety. Although the NTSB is aware that spectrum sharing within the 5-GHz band is needed for technological and economic growth, our comments stressed the importance of secure, protected communication frequencies for connected vehicle initiatives. Spectrum sharing could put the frequencies at risk of dangerous interference, and much is still unknown about frequency interference when it comes to vast numbers of connected vehicles in motion. A single

---

[140] See www.its.dot.gov/presentations/trb_2013/Safety_Pilot_TRB_files/frame.htm, accessed May 22, 2013.

[141] See ConnectedVehicleTechnologyFactSheet-081012[1]pdf, accessed May 22, 2013.

[142] See www.its.dot.gov/presentations/trb_2013/Safety_Pilot_TRB_files/frame.htm, accessed May 22, 2013.

[143] See 78 FR 21320, April 10, 2013.

incident, such as the case of interference encountered by the FAA with its Terminal Doppler Weather Radar, could stall progress and cause concern within the industry.[144]

The NTSB believes that initiating the connected vehicle technology rulemaking process would allow manufacturers to expedite and prioritize research and development of the new technology. The opportunity to improve transportation safety must not be delayed by issues associated with interference on the 5-GHz band. The NTSB concludes that the analysis of potential harmful interference from large numbers of U-NII devices is critical prior to opening up safety-sensitive frequencies to these devices, particularly to prevent delay in rulemaking on connected vehicle technologies in both passenger cars and commercial vehicles.

2.6 School Bus Occupant Protection

This collision sequence involved the truck hitting the school bus and the bus subsequently striking a traffic beacon support pole. The first impact caused a sudden lateral acceleration and change in direction of the bus (from forward to lateral and counterclockwise), and caused the occupants, especially those in the rear of the bus, to move laterally toward the truck. The second impact, on the right side of the bus, resulted in a rapid deceleration of the bus and caused the occupants who had moved toward the truck to then rapidly travel back toward the right (passenger) side of the bus.

Intrusion into the passenger compartment occurred in rows 8–10 on the left side of the bus (from the truck collision) and in rows 7–9 on the right side (from the traffic beacon support pole). The fatally and seriously injured passengers were seated in rows 7–11, in the area of the two side impacts, where they would have experienced higher impact forces and also higher rotational accelerations. The NTSB concludes that the fatally and severely injured passengers were seated in the back half of the school bus, in the area of the higher impact forces and accelerations.

Ten passengers sustained minor injuries, which consisted of muscle strains, contusions, and lacerations. The majority of these passengers were seated toward the front of the bus, in an area away from the intrusion zone and with lower accelerations.

The state of New Jersey requires that school buses of all sizes be equipped with seat belts, and that all bus passengers wear a properly adjusted and fastened seat belt or other child restraint system at all times while the bus is in operation.[145] The restraint system must meet applicable federal standards. The school bus was equipped with passenger lap belts at all seating positions. The NBCRSD transportation director reported that instruction on how to properly adjust the lap belt is reviewed with students yearly during emergency evacuation drills held at the beginning of the school year. However, the transportation director also stated that in the two weeks following the crash, she rode on the bus route and observed that the majority of students were wearing their seat belts but were not properly adjusting them. Postcrash examination of the

---

[144] The FAA issue is recounted in "Background," paragraph 8, of the subject NPRM. (See 78 FR 21322, April 10, 2013.)

[145] New Jersey state law does not hold either the owner or the operator of a school bus liable for failure to fasten or properly adjust a passenger seat belt or other child restraint system.

lap belts in the occupied seats of the school bus indicated that some belts had been worn but were not adjusted to fit properly. Furthermore, injuries to the fatally injured occupant and examination of the lap belts in her row indicated that she was likely unbelted. The NTSB concludes that some students on the school bus wore their lap belts improperly or not at all.

A properly worn lap belt can provide restraint for the lower body and reduce occupant motion out of the seating compartment, especially in lateral impacts and rollovers. Another school bus crash investigated by the NTSB—which occurred in Port St. Lucie, Florida, on March 26, 2012—also involved a severe lateral impact collision from a large mass vehicle (a truck-tractor semitrailer) hitting the right side of a school bus. In this crash, lap-belted occupants in the intrusion zone were initially displaced away from the impacting truck by the combined seat and restraint system, which may have reduced injuries from direct contact with intruding surfaces. Although the passengers' upper bodies were still free to rotate, the lap-belted passengers experienced reduced lateral excursion when compared to an unbelted passenger.

To further examine occupant kinematics for similar crash sequences, the NTSB conducted simulations of the Chesterfield and Port St. Lucie school bus crashes. The simulation results for Chesterfield indicated that unbelted occupants traveled from one side of the school bus to the other during the first and second impacts, with the truck and the traffic beacon support pole, respectively, causing significant potential for injury. In the case of Port St. Lucie, lap-belted occupants experienced reduced motion and less potential for injury from impacting other occupants and from high impact forces. Based on information from both the Chesterfield and the Port St. Lucie crashes, the NTSB concludes that lap belts can provide a benefit to passengers who wear them properly.

The simulations also showed that lap-belted occupants were still subject to injuries as a result of their flailing upper bodies. (See figure 23.) Chesterfield passengers reported that some injuries resulted from impact with the bus sidewall or interior components. Similarly, in the Port St. Lucie crash, passengers were documented as impacting the school bus sidewall. In a 1999 special report on bus crashworthiness, the NTSB addressed protection for school bus occupants and concluded that compartmentalization is incomplete in that it does not protect passengers during lateral impacts with vehicles of large mass or in rollovers[146] (NTSB 1999). Additional investigations found that exemptions of the rigid sidewalls, sidewall components, and seat frames from passenger protection standards may place occupants at risk during lateral impacts (NTSB 2002; 2001a; 2000).

---

[146] Compartmentalization is the current form of passenger protection in large school buses nationwide. Only six states require the installation of seat belts in large school buses. Compartmentalization is a passive system, requiring no action by the passenger, and functions by forming a compartment around the passenger with closely spaced, energy-absorbing seats that deform in a crash, allowing the passenger to "ride down" the collision. This system was designed to contain passengers within their seating compartments during frontal and rear impact collisions.

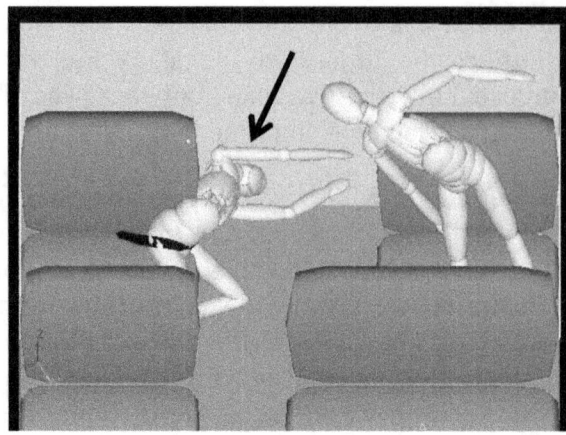

Figure 23. Still image of bus passengers from Chesterfield Madymo simulation, showing flailed upper body of lap-belted occupant (marked with black arrow).

Many school buses are currently designed with compartmentalization as the sole occupant protection system. Some other school buses are equipped with compartmentalization and lap belts. In either case, the lateral motion of unbelted passengers or the upper body flailing of lap-belted passengers exposes bus occupants to injury-producing components in lateral collisions. The NTSB concludes that adding protection to school bus sidewalls, sidewall components, seat frames, and other currently exempt interior components can reduce injuries to unbelted or lap-belted school bus passengers.

In 2001, the NTSB issued a recommendation to NHTSA to develop and incorporate into the FMVSS performance standards for school buses that address passenger protection for sidewalls, sidewall components, and seat frames (Safety Recommendation H-01-40). According to NHTSA, it is conducting ongoing research to quantify the magnitude of the injury problem to school bus occupants during a side impact, and to evaluate potential methods for mitigating these injuries. NHTSA reports that crash data analysis is continuing to better define the conditions and locations of head impacts in bus crashes. NHTSA completed school bus side impact sled testing in 2011 (at the Vehicle Research Testing Center), resulting in an evaluated baseline for injury with no padding and with padded interior surfaces. The sled testing demonstrated a reduction in injury values through the added padding. The agency is contracting with the Mercer University Engineering Research Center in a joint research effort to: (1) develop a finite element model of a typical school bus construction, and (2) study the effects on occupant protection of various levels and types of padding added to the bus sidewall or roof area (Elias, Sullivan, and McCray 2003).

Safety Recommendation H-01-40 is currently classified "Open—Acceptable Response." However, because ensuring school bus safety is a high priority, the NTSB is reiterating this recommendation and changing the classification to "Open—Unacceptable Response" due to NHTSA's delay in addressing these interior school bus passenger protections.

The Port St. Lucie accident onboard VER system provided a unique opportunity to study lap-belted occupant motion in a severe school bus lateral impact. The fatally injured passenger in this crash was seated aft of the area of intrusion on the opposite side of the impact. Although he was not directly visible in any of the camera views, the motion of nearby passengers in a similar orientation relative to the crash forces and postcrash photographs provided insight into his dynamics. This passenger likely experienced the highest accelerations due to his row 10 seating position and also the greatest flail due to his aisle seating position on the opposite side of the impact. Further, physical evidence on the lap belt and injuries to this passenger indicated that he was initially properly restrained with the lap belt. Unfortunately, at some point during the crash sequence, the integrity of the seating system was lost. A combination of high impact forces and high angular accelerations resulting from the side impact and the dynamic motion of the belted passenger likely caused the seat pan to become detached from the seat frame.[147] The NTSB concludes that in the Port St. Lucie crash, the combination of high forces, lack of upper body restraint, and loss of seating system integrity resulted in fatal injuries to one passenger.

The entire seating system is a critical component when evaluating occupant motion in a school bus equipped with passenger restraints. The motion of an occupant in the seat and the occupant's interaction with the seat and the restraints are important factors in determining the type, mechanism, and potential severity of any resulting injury (Elias, Sullivan, and McCray 2003). In the Port St. Lucie crash, the fatally injured passenger was seated at a location where the seat pan separated from the seat frame, reducing the effectiveness of the combined seating and restraint system and likely contributing to his injuries.

As a result, the NTSB concludes that the design of a school bus restraint system must also focus on maintaining the integrity of the seating system during severe impact scenarios. Although there is no federal mandate for restraints in large school buses, many schools are purchasing buses equipped with restraint systems, and school bus seating systems need to be addressed as a whole to prevent unintended injuries from a failure of one component in the system. The NTSB recommends that the School Bus Manufacturers Technical Council (SBMTC)[148] develop a recommended practice for establishing and safeguarding the structural integrity of the entire school bus seating and restraint system, including the seat pan attachment to the seat frame, in severe crashes—in particular, those involving lateral impacts with vehicles of large mass.

A well designed seating and restraint system in school buses is only effective if used properly. Ensuring proper belt use is challenging but critical to the performance of seat belts in large school buses (NHTSA 2002). As representative of the problem, a University of Alabama study found that the average seat belt use over a two-year period was only 61.5 percent (Turner and others 2010; NTSB 2009a).

The NTSB investigated a school bus run-off-the-road and rollover crash in Milton, Florida, in which the school bus was equipped with lap belts for each seating position (NTSB

---

[147] Although the Chesterfield school bus was also manufactured by IC Bus, LLC, and also experienced lateral forces and angular accelerations, the seat pans did not detach from the seat frames, as occurred in the Port St. Lucie crash.

[148] The SBMTC, an organization within the National Association of State Directors of Pupil Transportation Services (NASDPTS), was established in 1995, as a subsidiary of the NASDPTS Supplier Council, to function as the industry's technical advisor. The council provides a forum for members to address technical and government-related issues concerning the manufacture and acceptability of school bus chassis and school bus bodies. See also www.nasdpts.org/SBMTC/, accessed May 15, 2013.

2009b). Even with a chaperone on the school bus, the NTSB found that some students either loosened or unbuckled the lap belt after the chaperone's inspection. As stated earlier, following the Chesterfield crash, when the NBCRSD transportation director rode the same bus route, she also noticed that many students were not properly wearing their lap belts. The NTSB concludes that better student, parent, and school district education and training may increase usage and the proper fit of passenger seat belts in school buses.

Because of the potential for flailing injuries with lap-only belts, as shown in figure 23, the NTSB simulations also examined the injury potential for occupants restrained with lap and shoulder belts. The results for both the Chesterfield and the Port St. Lucie simulations found that lap- and shoulder-belted occupants were better protected due to the reduced flail of the upper body. The NTSB concludes that in severe side impact crashes, such as Chesterfield and Port St. Lucie, properly worn lap and shoulder belts reduce injuries related to upper body flailing commonly seen with lap belts only and, therefore, provide the best protection for school bus passengers.

There are several school bus, seating, and seat belt manufacturers, and a myriad of options are available to school districts—each requiring that the students, parents, school bus drivers, and others in charge of pupil transportation be knowledgeable about seat belt systems. The NTSB, therefore, recommends that NASDPTS, the National Association for Pupil Transportation (NAPT), the National School Transportation Association (NSTA), the SBMTC, and the National Safety Council, School Transportation Section, develop guidelines and include them in the next update of the National School Transportation Specifications and Procedures to assist schools in training bus drivers, students, and parents on the importance and proper use of school bus seat belts, including manual lap belts, adjustable lap and shoulder belts, and flexible seating systems.

The NTSB further recommends that the states of California, Florida, Louisiana, New Jersey, New York, and Texas—which currently require school bus passenger seat belts—develop: (1) a handout for their school districts to distribute annually to students and parents about the importance of the proper use of all types of passenger seat belts on school buses, including the potential harm of not wearing a seat belt or wearing one but not adjusting it properly; and (2) training procedures for schools to follow during the twice yearly emergency drills to show students how to wear their seat belts properly. In addition, the NTSB recommends that the states of California, Florida, Louisiana, New Jersey, New York, and Texas—upon publication of the National School Transportation Specifications and Procedures document—revise the handout and training procedures developed above to align with the national procedures as appropriate.

In the 1999 school bus crashworthiness report, the NTSB concluded that: (1) because of compartmentalization, school bus passenger seats are safer now than they were before 1977; and (2) during lateral impact with vehicles of large mass and in rollovers, passengers do not always remain completely within the seating compartment (NTSB 1999). Based on the Chesterfield and Port St. Lucie investigations, the NTSB concludes that though school buses are extremely safe, properly worn passenger seat belts make the school bus safer, especially in severe side impacts and rollovers.

Because there is no federal requirement for every school bus to have seat belts, each state must decide whether to install lap and shoulder or lap-only belts on their school buses. The NTSB recognizes that transporting students on school buses is critical to maintaining the safety of students and strongly discourages steps to transport children by other means (NHTSA 2013). Therefore, the NTSB recommends that NASDPTS, NAPT, and NSTA provide their members with educational materials on lap and shoulder belts providing the highest level of protection for school bus passengers, and advise states or school districts to consider this added safety benefit when purchasing seat belt-equipped school buses.

3 Conclusions

3.1 Findings

1. None of the following were factors in the crash: (1) alcohol impairment or illicit drug use by the school bus driver, or alcohol, over-the-counter, prescription medication, or illicit drug use by the truck driver; (2) in-vehicle or external distractions, including cell phone use; (3) truck driver fatigue; (4) operations by Garden State Transport Corporation, the school bus motor carrier; (5) school bus mechanical defects or deficiencies; or (6) weather.

2. The emergency response was timely and adequate.

3. The truck was within the school bus driver's available line of sight and within a hazardous proximity when the bus driver began to cross the intersection.

4. The driver of the truck was driving in excess of the posted speed limit before braking for the impending collision, and this higher speed contributed to the severity of the crash.

5. The school bus driver did not effectively scan Burlington County Road 528 for oncoming traffic and failed to observe the approaching truck prior to impact.

6. The school bus driver was fatigued due to acute sleep loss, chronic sleep debt, and poor sleep quality associated with his medical conditions and alcohol use; the sedative side effects from prescription medications; and the synergistic effect of these factors.

7. The school bus driver's fatigue contributed to his reduced vigilance and detection of the approaching truck.

8. The school bus driver failed to disclose pertinent information about his medical history as required on the commercial driver's license medical certification examination form, which prevented the accurate assessment of his qualifications to drive a school bus in commercial operations.

9. The commercial driver's license medical examiner did not thoroughly evaluate the school bus driver for medical conditions that could have disqualified him from receiving a commercial driver's license.

10. Based on the school bus driver's combination of medical conditions and use of multiple prescription medications, it is likely that he would not have been medically certified to drive a school bus if: (1) he had fully disclosed his medical history on the commercial driver's license medical certification examination form, or (2) the medical examiner had completed a more thorough evaluation.

11. Some medical professionals who are authorized to perform medical examinations and certify commercial drivers as fit to drive may lack the knowledge and information critical to certification decisions; consequently, drivers with serious medical conditions may not be sufficiently evaluated to determine whether their conditions pose a risk to highway safety.

12. The combination of the truck's defective brakes and overweight condition reduced its overall braking efficiency, thereby contributing to the severity of the crash.

13. Had the truck been equipped with an onboard weighing system, the truck driver would have had information about its overweight condition.

14. Had the truck been equipped with an onboard brake stroke monitoring system, the truck driver could have had information about the out-of-adjustment and impending out-of-adjustment brakes.

15. Due to improper installation of the lift axle brake system by Automated Waste Equipment, the brakes on the truck would not likely have met FMVSS 121 timing requirements at the time of manufacture.

16. Had the lift axle brake system been properly installed on the truck, air would have been applied to the brakes earlier, thereby reducing the severity of the crash.

17. The driver of the truck did not ensure that his vehicle was within allowable weight restrictions, despite the requirement to do so and the fact that there was a weigh station located 2.5 miles from his originating point.

18. Herman's Trucking Inc. did not have effective oversight of its drivers to ensure that the company truck was routed, as required by law, to the closest weighing scales, thereby preventing the transportation of overweight loads on public roads.

19. Although the sight distance available for vehicles stopping at the stop line on northbound Burlington County Road 660 was inadequate, it did not contribute to the cause of the crash because the school bus driver stopped forward of the stop line, where there was sufficient sight distance.

20. The skew of the Burlington County Road 528–660 intersection required the school bus driver to turn farther to the left to observe oncoming traffic, which called for more continuous scanning to safely cross the highway.

21. Burlington County took direct steps following the crash to improve the safety of the Burlington County Road 528–660 intersection.

22. Connected vehicle technology could have provided active warnings to the school bus driver of the approaching truck and possibly prevented the crash.

23. The analysis of potential harmful interference from large numbers of unlicensed national information infrastructure devices is critical prior to opening up

safety-sensitive frequencies to these devices, particularly to prevent delay in rulemaking on connected vehicle technologies in both passenger cars and commercial vehicles.

24. The fatally and severely injured passengers were seated in the back half of the school bus, in the area of the higher impact forces and accelerations.

25. Some students on the school bus wore their lap belts improperly or not at all.

26. Based on information from both the Chesterfield, New Jersey, and the Port St. Lucie, Florida, crashes, lap belts can provide a benefit to passengers who wear them properly.

27. Adding protection to school bus sidewalls, sidewall components, seat frames, and other currently exempt interior components can reduce injuries to unbelted or lap-belted school bus passengers.

28. In the Port St. Lucie, Florida, crash, the combination of high forces, lack of upper body restraint, and loss of seating system integrity resulted in fatal injuries to one passenger.

29. The design of a school bus restraint system must also focus on maintaining the integrity of the seating system during severe impact scenarios.

30. Better student, parent, and school district education and training may increase usage and the proper fit of passenger seat belts in school buses.

31. In severe side impact crashes, such as Chesterfield, New Jersey, and Port St. Lucie, Florida, properly worn lap and shoulder belts reduce injuries related to upper body flailing commonly seen with lap belts only and, therefore, provide the best protection for school bus passengers.

32. Although school buses are extremely safe, properly worn passenger seat belts make the school bus safer, especially in severe side impacts and rollovers.

3.2 Probable Cause

The National Transportation Safety Board determines that the probable cause of the Chesterfield, New Jersey, crash was the school bus driver's failure to observe the Mack roll-off truck, which was approaching the intersection within a hazardous proximity. Contributing to the school bus driver's reduced vigilance were cognitive decrements due to fatigue as a result of acute sleep loss, chronic sleep debt, and poor sleep quality, in combination with, and exacerbated by, sedative side effects from his use of prescription medications. Contributing to the severity of the crash was the truck driver's operation of his vehicle in excess of the posted speed limit, in addition to his failure to ensure that the weight of the vehicle was within allowable operating restrictions. Further contributing to the severity of the crash were the defective brakes on the truck and its overweight condition due to poor vehicle oversight by Herman's Trucking, along with improper installation of the lift axle brake system by the final stage manufacturer—all of which degraded the truck's braking performance. Contributing to the severity of passenger injuries were the nonuse or misuse of school bus passenger lap belts; the lack of passenger protection from interior sidewalls, sidewall components, and seat frames; and the high lateral and rotational forces in the back portion of the bus.

4 Recommendations

As a result of its investigation, the National Transportation Safety Board makes the following safety recommendations:

4.1 New Recommendations

To the Federal Motor Carrier Safety Administration:

Require that all persons applying for inclusion on the National Registry of Certified Medical Examiners have both a thorough knowledge of pharmacology and current prescribing authority. (H-13-27)

To the National Highway Traffic Safety Administration:

Develop minimum performance standards for onboard vehicle weighing systems for trucks that have a gross vehicle weight rating of 10,000 pounds or more and are typically field loaded and used in the transportation of aggregates or earthen construction materials, raw natural resources, and garbage or refuse, or in logging and timber operations, or agricultural operations. (H-13-28)

Once minimum performance standards for onboard vehicle weighing systems are established, require these systems to be installed on newly manufactured trucks that have a gross vehicle weight rating of 10,000 pounds or more and are typically field loaded and used in the transportation of aggregates or earthen construction materials, raw natural resources, and garbage or refuse, or in logging and timber operations, or agricultural operations. (H-13-29)

Develop minimum performance standards for connected vehicle technology for all highway vehicles. (H-13-30)

Once minimum performance standards for connected vehicle technology are developed, require this technology to be installed on all newly manufactured highway vehicles. (H-13-31)

To the states of California, Florida, Louisiana, New Jersey, New York, and Texas:

Develop: (1) a handout for your school districts to distribute annually to students and parents about the importance of the proper use of all types of passenger seat belts on school buses, including the potential harm of not wearing a seat belt or wearing one but not adjusting it properly; and (2) training procedures for schools to follow during the twice yearly emergency drills to show students how to wear their seat belts properly. (H-13-32)

Upon publication of the National School Transportation Specifications and Procedures document, revise the handout and training procedures developed in Safety Recommendation H-13-32 to align with the national procedures as appropriate. (H-13-33)

To the National Truck Equipment Association:

Notify your members of the Chesterfield crash and of the need to check their vehicles for potentially improper lift axle brake installation. (H-13-34)

To the National Association of State Directors of Pupil Transportation Services, National Association for Pupil Transportation, National School Transportation Association, School Bus Manufacturers Technical Council, and National Safety Council, School Transportation Section:

Develop guidelines and include them in the next update of the National School Transportation Specifications and Procedures to assist schools in training bus drivers, students, and parents on the importance and proper use of school bus seat belts, including manual lap belts, adjustable lap and shoulder belts, and flexible seating systems. (H-13-35)

To the National Association of State Directors of Pupil Transportation Services, National Association for Pupil Transportation, and National School Transportation Association:

Provide your members with educational materials on lap and shoulder belts providing the highest level of protection for school bus passengers, and advise states or school districts to consider this added safety benefit when purchasing seat belt-equipped school buses. (H-13-36)

To the School Bus Manufacturers Technical Council:

Develop a recommended practice for establishing and safeguarding the structural integrity of the entire school bus seating and restraint system, including the seat pan attachment to the seat frame, in severe crashes—in particular, those involving lateral impacts with vehicles of large mass. (H-13-37)

To Herman's Trucking Inc.:

Develop and implement route procedures to prevent the transportation of overweight loads. (H-13-38)

4.2 Previously Issued Recommendations Reiterated in This Report

The National Transportation Safety Board also reiterates the following safety recommendations:

To the Federal Motor Carrier Safety Administration:

Develop a comprehensive medical oversight program for interstate commercial drivers that contains the following program elements:

Individuals performing medical examinations for drivers are qualified to do so and are educated about occupational issues for drivers. (H-01-17)

Medical certification regulations are updated periodically to permit trained examiners to clearly determine whether drivers with common medical conditions should be issued a medical certificate. (H-01-19)

Individuals performing examinations have specific guidance and a readily identifiable source of information for questions on such examinations. (H-01-20)

The review process prevents, or identifies and corrects, the inappropriate issuance of medical certification. (H-01-21)

To the National Highway Traffic Safety Administration:

Develop minimum performance standards for onboard brake stroke monitoring systems for all air-braked commercial vehicles. (H-12-58)

Once the performance standards in Safety Recommendation H-12-58 have been developed, require that all newly manufactured air-braked commercial vehicles be equipped with onboard brake stroke monitoring systems. (H-12-59)

4.3 Previously Issued Recommendation Reiterated and Reclassified in This Report

The National Transportation Safety Board reiterates and reclassifies the following safety recommendation:

To the National Highway Traffic Safety Administration:

Develop and incorporate into the FMVSS, performance standards for school buses that address passenger protection for sidewalls, sidewall components, and seat frames. (H-01-40)

Safety Recommendation H-01-40 is reiterated and reclassified "Open—Unacceptable Response."

BY THE NATIONAL TRANSPORTATION SAFETY BOARD

DEBORAH A. P. HERSMAN
Chairman

ROBERT L. SUMWALT
Member

CHRISTOPHER A. HART
Vice Chairman

MARK R. ROSEKIND
Member

EARL F. WEENER
Member

Adopted: July 23, 2013

References

AASHTO (American Association of State Highway and Transportation Officials). 2011. *A Policy on Geometric Design of Highways and Streets*, 6th edition. Washington, DC: AASHTO.

Babkoff, H., T. Caspy, and M. Mikulincer. 1991. "Subjective Sleepiness Ratings: The Effects of Sleep Deprivation, Circadian Rhythmicity and Cognitive Performance." *Sleep* 14: 534-539.

Belenky, G., N. J. Wesensten, D. R. Thorne, M. L. Thomas, H. C. Sing, D. P. Redmond, M. B. Russo, and T. J. Balkin. 2003. "Patterns of Performance Degradation and Restoration During Sleep Restriction and Subsequent Recovery: A Sleep Dose-Response Study." *Journal of Sleep Research* 12: 1–12.

Blumenthal, R., J. Braunstein, H. Connolly, A. Epstein, B. J. Gersh, and E. H. Wittels. 2002. *Cardiovascular Advisory Panel Guidelines for the Medical Examination of Commercial Motor Vehicle Drivers*, Report No. FMCSA-MCP-02-002. Washington, DC: FMCSA.

Bonnet, M. H. 2005. "Acute Sleep Deprivation." In *Principles and Practice of Sleep Medicine*, edited by M. Kryger, T. Roth, and W. Dement. Philadelphia, Pennsylvania: Elsevier-Sanders.

Carskadon, M., and W. Dement. 2005. "Normal Human Sleep: An Overview." In *Principles and Practice of Sleep Medicine*, edited by M. Kryger, T. Roth, and W. Dement. Philadelphia, Pennsylvania: Elsevier-Sanders.

Dinges, D. F. 1995. "Performance Effects of Fatigue." *Proceedings of the NTSB/NASA Fatigue Symposium*, 41–46. Washington, DC: NTSB.

Dinges, D. F., and N. Kribbs. 1991. "Performing While Sleepy: Effects of Experimentally Induced Sleepiness." In *Sleep, Sleepiness, and Performance*, edited by T. Monk, 98–128. Chichester, UK: John Wiley & Sons.

Dinges, D. F., F. Pack, K. Williams, K. A. Gillen, J. W. Powell, G. E. Ott, D. Aptowicz, and A. Pack. 1997. "Cumulative Sleepiness, Mood Disturbance, and Psychomotor Vigilance Performance Decrements During a Week of Sleep Restricted to 4–5 Hours per Night." *Sleep* 20: 267–277.

Drake, C. L., T. A. Roehrs, E. Burduvali, A. Bonahoom, M. Rosekind, and T. Roth. 2001. "Effects of Rapid Versus Slow Accumulation of Eight Hours of Sleep Loss." *Psychophysiology* 38: 979–987.

Elias, J. C., L. K. Sullivan, and L. B. McCray. 2003. *Large School Bus Safety Restraint Evaluation, Phase II*. Transportation Research Center, Inc., and Vehicle Research Testing Center. Washington, DC: NHTSA.

FMCSA (Federal Motor Carrier Safety Administration). 2007. *The Large Truck Crash Causation Study*, Technical Report No. FMCSA-RRA-07-017. Washington, DC: FMCSA.

FHWA (Federal Highway Administration). 2009. *Manual on Uniform Traffic Control Devices*. Washington, DC: FHWA.

————. 2000. *Roundabouts: an Informational Guide*, Report No. FHWA-RD-00-067. Washington, DC: FHWA.

Gaines, R. W., and W. G. Humphrey. 1993. "Spondylolisthesis." In *Operative Orthopaedics*, edited by M. W. Chapman. Philadelphia, Pennsylvania: Lippincott.

Galera, C., L. Orriols, K. M'Bailara, M. Laborey, B. Contrand, R. Ribereau-Gayon, F. Masson, S. Bakiri, C. Gabaude, A. Fort, B. Maury, C. Lemercier, M. Cours, M. Bouvard, and E. Lagarde. 2012. "Mind Wandering and Driving: Responsibility Case-Control Study." *BMJ* 345: e8105.

GAO (US Government Accountability Office). 2012. *Highway Safety: Selected Cases of Commercial Drivers With Potentially Disqualifying Impairments*, Report No. GAO-13-13. Washington, DC: GAO.

Gattis, J. L., and Sonny T. Low. 1997. *Intersection Angles and Drivers Field of View*. Fayetteville, Arkansas: Mack-Blackwell Rural Transportation Center, University of Arkansas.

Glenville, M., R. Broughton, A. M. Wing, and R. T. Wilkinson. 1978. "Effects of Sleep Deprivation on Short Duration Performance Measures Compared to the Wilkinson Auditory Vigilance Task." *Sleep* 1: 169–176.

Goel, N., H. Rao, J. S. Durmer, and D. F. Dinges. 2009. "Neurocognitive Consequences of Sleep Deprivation." *Seminars in Neurology* 29 (4): 320–339.

Gordon, C. 2007. "Driver Distraction Related Crashes in New Zealand." In *Distracted Driving*, edited by I. J. Faulks, M. Regan, M. Stevenson, J. Brown, A. Porter, and J. D. Irwin. 299–328. Sidney, NSW: Australasian College of Road Safety.

Harris, M. 2000. "Psychiatric Conditions With Relevance to Fitness to Drive." *Advances in Psychiatric Treatment* 6: 261–269.

Herner, B., D. Smedvy, and L. Ysander. 1966. "Sudden Illness as a Cause of Motor Vehicle Accidents." *British Journal of Industrial Medicine* 23: 37–41.

Heusser, R. B. 1991. "Heavy Truck Deceleration Rates as a Function of Brake Adjustment." *SAE International Congress & Exposition, February 25, 1991, Detroit, Michigan*, SAE Paper No. 910126 (Revision B). Warrendale, Pennsylvania: SAE International.

Howard, M. E., A. V. Desai, R. R. Grunstein, C. Hukins, J. G. Armstrong, D. Joffe, P. Swann, D. A. Campbell, and R. J. Pierce. 2004. "Sleepiness, Sleep-Disordered Breathing and Accident Risk Factors in Commercial Vehicle Drivers." *American Journal of Respiratory Critical Care Medicine* 170 (9): 1014–1021.

Hummel, T., S. Roscher, E. Pauli, M. Frank, J. Liefhold, W. Fleischer, and G. Kobal. 1996. "Assessment of Analgesia in Man: Tramadol Controlled Release Formula vs. Tramadol Standard Formulation." *European Journal of Clinical Pharmacology* 51: 31–38.

Kleitman, N. ed. 1963. "Deprivation of Sleep." *Sleep and Wakefulness* 215–229. Chicago, Illinois: University of Chicago Press.

Kreuger, G. P., and H. M. Leaman. 2011. "Commercial Truck and Bus Safety." *CTBSSP Synthesis 19: Effects of Psychoactive Chemicals on Commercial Driver Health Performance: Stimulants, Hypnotics, Nutritional, and Other Supplements. A Synthesis of Safety Practice*. Washington, DC: The National Academies, Transportation Research Board.

Kroemer, K. H. E., H. J. Kroemer, and K. E. Kroemer-Elbert. 1990. *Engineering Physiology: Bases of Human Factors/Ergonomics*. New York: Van Nostrand Reinhold.

Lamond, N., and D. Dawson. 1999. "Quantifying the Performance Impairment Associated With Fatigue." *Journal of Sleep Research* 8: 255–262.

Land Transport NZ (Land Transport New Zealand). 2005. *Road Safety Issues*. Wellington, New Zealand: Land Transport NZ.

Mack (Mack Trucks, Inc.). 2003. *Body Installer's Guide for Mack Class 8 Chassis*. Allentown, Pennsylvania: Mack Trucks, Inc.

NHTSA (National Highway Traffic Safety Administration). 2013. *Traffic Safety Facts, 2002-2011 Data, School Transportation-Related Crashes*, DOT HS 811 746. Washington, DC: NHTSA.

———. 2002. *School Bus Safety: Crashworthiness Research*, Report to Congress. Washington, DC: NHTSA.

NIAAA (National Institute on Alcohol Abuse and Alcoholism). 1998. *Alcohol and Sleep*, Alcohol Alert No. 41. Washington, DC: NIAAA.

NJDOT (New Jersey Department of Transportation). 2002. *Roadway Design Manual*. Trenton, New Jersey: NJDOT.

NTIA (National Telecommunications and Information Administration). 2013. *NTIA Evaluation of the 5350–5470 MHz and 5850–5925 MHz Bands Pursuant to Section 6406(b) of the Middle Class Tax Relief and Job Creation Act of 2012*. Washington, DC: US Department of Commerce.

NTSB (National Transportation Safety Board). 2012. *Highway–Railroad Grade Crossing Collision, US Highway 95, Miriam, Nevada, June 24, 2011*, NTSB/HAR-12/03. Washington, DC: NTSB.

———. 2009a. *School Bus Bridge Override Following Collision With Passenger Vehicle, Interstate Highway 565, Huntsville, Alabama, November 20, 2006*, NTSB/HAB-09/02. Washington, DC: NTSB.

———. 2009b. *School Bus Loss of Control and Rollover, Interstate 10, Milton, Florida, May 28, 2008*, NTSB/HAB-09/03. Washington, DC: NTSB.

———. 2006. *Collision Between a Ford Dump Truck and Four Passenger Cars, Glen Rock, Pennsylvania, April 11, 2003*, NTSB/HAR-06/01. Washington, DC: NTSB.

———. 2002. *Collision Between Truck-Tractor Semitrailer and School Bus Near Mountainburg, Arkansas, May 31, 2001*, NTSB/HAR-02/03. Washington, DC: NTSB.

———. 2001a. *Collision of CSXT Freight Train and Murray County School District Bus Near Conasauga, Tennessee, March 28, 2000*, NTSB/HAR-01/03. Washington, DC: NTSB.

———. 2001b. *Motorcoach Run-Off-the-Road Accident, New Orleans, Louisiana, May 9, 1999*, NTSB/HAR-01/01. Washington, DC: NTSB.

———. 2000. *School Bus and Dump Truck Collision Near Central Bridge, New York, October 21, 1999*, NTSB/HAR-00/02. Washington, DC: NTSB.

———. 1999. *Bus Crashworthiness*, NTSB/SIR-99/04. Washington, DC: NTSB.

Rayman, R. B., J. D. Hastings, W. B. Kruyer, and R. A. Levy. 2001. *Clinical Aviation Medicine*. New York: Castle Connolly.

Rosenblatt, R. C. 2004. "Grieving While Driving." *Death Studies* 28 (7): 679–686.

Rumar, K. 1990. "The Basic Driver Error: Late Detection." *Ergonomics* 33: 1281–1290.

Stutts, J. C., J. W. Wilkins, O. J. Scott, and B. V. Vaughn. 2003. "Driver Risk Factors for Sleep-Related Crashes." *Accident Analysis & Prevention* 35: 321–331.

Turner, D. S., J. K. Lindly, E. Tedla, K. Anderson, and D. Brown. 2010. *Summary Report, Alabama School Bus Seat Belt Pilot Project*, 07407-1. Tuscaloosa, Alabama: University Transportation Center for Alabama.

Van Dongen, H. P. A., G. Maislin, J. M. Mullington, and D. F. Dinges. 2003. "The Cumulative Cost of Additional Wakefulness: Dose Response Effects on Neurobehavioral Functions and Sleep Physiology From Chronic Sleep Restriction and Total Sleep Deprivation." *Sleep* 26: 117–126.

Wilkinson, R. T., R. S. Edwards, and E. Haines. 1966. "Performance Following a Night of Reduced Sleep." *Psychonomic Science* 5: 471–472.

Appendix A: Investigation

The National Transportation Safety Board was notified of this crash on February 16, 2012. An investigative team was dispatched with members from the Washington, D.C.; Gardena, California; and Arlington, Texas, offices. Groups were established to investigate human performance; motor carrier operations; onboard recorders; and highway, vehicle, and survival factors.

Parties to the investigation were representatives from the Federal Motor Carrier Safety Administration, National Highway Traffic Safety Administration, New Jersey State Police, State of New Jersey Motor Vehicle Commission School Bus Inspection Unit, Burlington County Prosecutors Office, Burlington County Engineers Office, Chesterfield Township Police Department, Florence Township Police Department, Navistar, Inc. (IC Bus, LLC), Mack Trucks, Inc., Bendix Commercial Vehicle Systems, LLC, Garden State Transport, Inc., Palfinger American Roll-off, and Herman's Trucking Inc.

No depositions were taken, and no public hearing was held.

Appendix B: Data From Truck's Garmin GPS Device

This appendix provides information from the truck driver's Garmin nüvi 1390 device. Because the device does not record global positioning system (GPS) accuracy, an industry-typical error diameter of 15 meters (approximately 49.2 feet) was used to approximate the precision of the GPS data. Table B-1 summarizes the uncertainty of the speed calculations based on this approximation.

Table B-1. GPS data from truck, February 16, 2012.

Time (UTC)[a]	Latitude (°)	Longitude (°)	Time Interval (seconds)	Distance Traveled (feet)	Average Speed (mph)	Uncertainty (+/- mph)
13:14:59	40.134606	-74.668713				
			17	1178	47.3	2.0
13:15:16	40.132617	-74.665385				
			14	1030	50.2	2.4
13:15:30	40.130896	-74.662456				
			5	345	47.0	6.7
13:15:35	40.130315	-74.66148				
			1	60	49.8	33.5
13:15:36	40.130214	-74.661311				
[a] UTC = Coordinated Universal Time.						

The last point reported at 13:15:36 is not considered to be a precise recording of the vehicle's position. After crossing the Burlington County Road 528–660 intersection between 13:15:30 and 13:15:35, the truck exited the roadway. The Garmin GPS device is not capable of recording data outside the roadway, so this last position is not an exact recording of the vehicle's position.

Appendix C: NTSB Sight Distance Tests

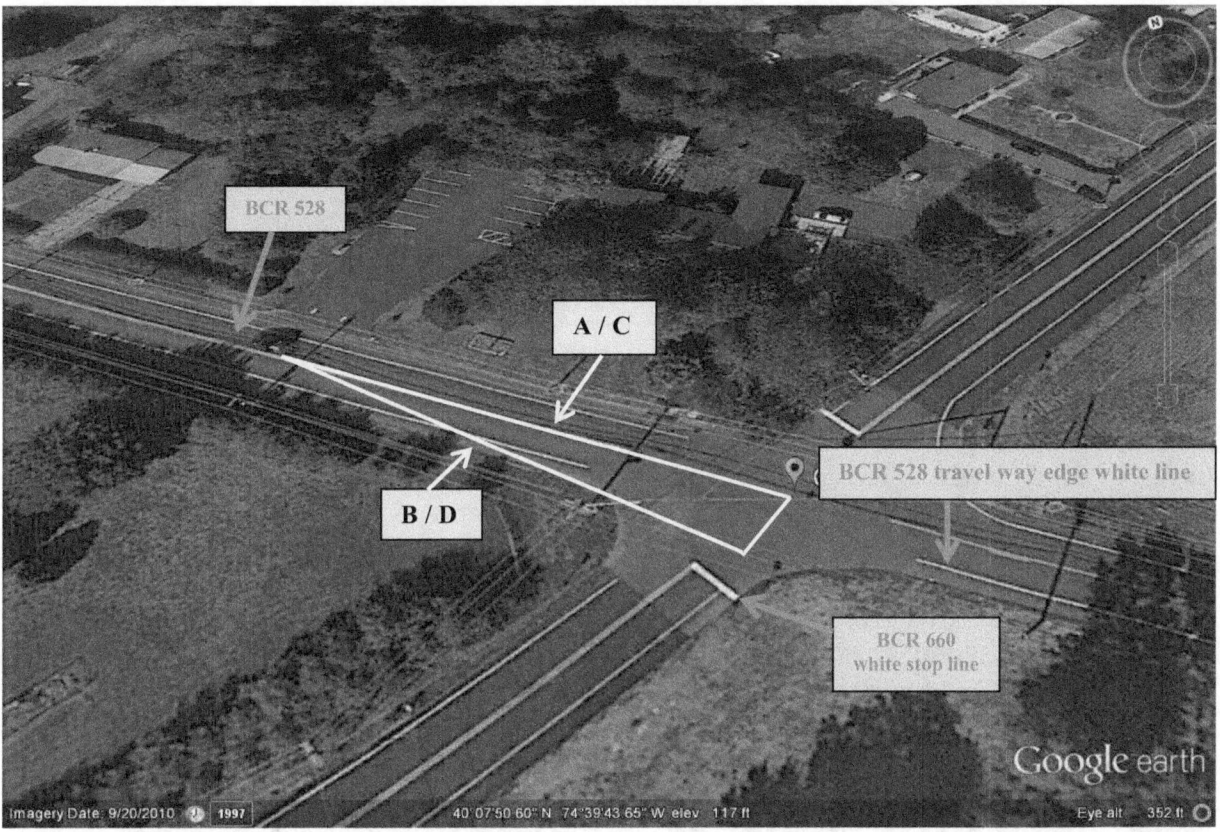

Figure C-1. Aerial view of BCR 528–660 intersection showing parameters for sight distance tests, keyed to table C-1. (Google imagery, September 20, 2010)

Table C-1. Variables for sight distance testing from BCR 660.

	1/30 (feet)	2/25 (feet)	5/16 (feet)	7/10 (feet)
A	287.24	345.11	686.34	814.76
B	274.69	335.28	680.69	811.17
C	206.36	242.74	463.13	782.69
D	195.66	233.29	459.78	787.27

A: Sight distance from truck driver's view to intersection of edge line of eastbound approach and extended centerline of northbound approach.
B: Clear sight triangle (distance from front bumper of school bus to front of truck).
C: Sight distance from bus driver's point of view to intersection of edge line of eastbound approach and extended centerline of northbound approach.
D: Clear sight triangle (distance from bus driver's eye to front of truck).

The starting point (location 1) for the first sight triangle was the existing stop line, which measured 30 feet from the edge of the travel way. Location 2 was 25 feet from the edge of the travel way and met the criteria established in *New Jersey Statutes*, section 39:4-144, which states that a vehicle shall stop within 5 feet of a stop line. Location 5 was approximately where the school bus driver stated that he stopped, and was measured at 16 feet from the edge of the travel way (and 14 feet past the white stop line). The final exercise (location 7) was conducted at a point 10 feet from the edge of the travel way (and 20 feet past the white stop line) to establish the minimum American Association of State Highway and Transportation Officials (AASHTO) sight triangle. It should be noted that the exercise conducted at location 7 also met the minimum requirements of the *Manual on Uniform Traffic Control Devices* (MUTCD; section 3B.16) for the placement of a stop line in relationship to the extended curb line or edge of pavement (FHWA 2009), and the driver's eye was above the 18-foot AASHTO decision point. At location 7, the driver had to turn approximately 110 degrees to view the oncoming truck.

Sight triangles were measured for the northbound approach of Burlington County Road (BCR) 660 at various increments from the edge line of the eastbound travel lane in accordance with AASHTO and MUTCD standards for a single unit truck, which is the design vehicle length most closely related to a school bus (AASHTO 2011, 2-3, table 2-1a).

Figure 6-A from the *Roadway Design Manual* (NJDOT 2002):

SIGHT DISTANCE AT INTERSECTIONS FOR LEFT, OR RIGHT TURNING & CROSSING VEHICLES WITH STOP CONTROL

FIGURE: 6-A

BDC07MR-05

Design Speed	Left-Turn			Right-Turn or Cross		
	P	**SU**	**WB**	**P**	**SU**	**WB**
25	280	350	425	240	315	385
30	335	420	510	290	375	465
35	390	490	595	335	440	540
40	445	560	680	385	500	620
45	500	630	760	430	565	695
50	555	700	845	480	625	775
55	610	770	930	530	690	850
60	665	840	1015	575	750	930
65	720	910	1100	625	815	1005
70	775	980	1185	670	875	1085

Intersection Sight Distance(d) Stop Control on Minor Road Two Lane Highway

For highways with more than 2 lanes or when approach grade on minor road exceeds 3%, the distance (d) must be calculated using the formula: $d = 1.47 V t_g$

Design Vehicle	Time Gap, t_g Left-Turn	Time Gap, t_g Right-Turn & Cross
P	7.5 (See Notes)	6.5 (See Notes)
SU	9.5 (See Notes)	8.5 (See Notes)
WB	11.5 (See Notes)	10.5 (See Notes)

Notes: 1. For left turn or crossing add 0.5 sec. for P and 0.7 sec. for SU & WB for each additional lane crossed.

2. For each percent the upgrade on minor road exceeds 3%, add 0.1 sec for right turn or crossing and 0.2 sec for left turn

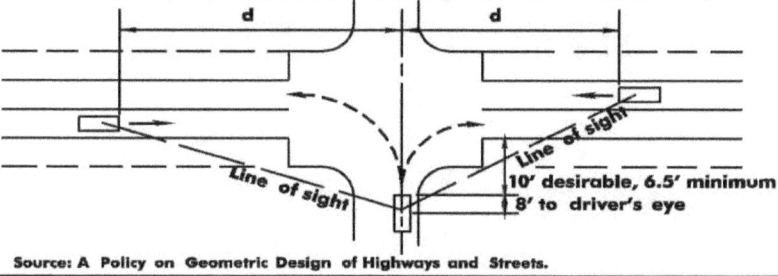

10' desirable, 6.5' minimum
8' to driver's eye

Source: A Policy on Geometric Design of Highways and Streets.

www.ingramcontent.com/pod-product-compliance
Lightning Source LLC
Chambersburg PA
CBHW080304180526
45167CB00006B/2657